Hermann Bähr

Orbital Effects in Spaceborne Synthetic Aperture Radar Interferometry

Karlsruher Institut für Technologie
Schriftenreihe des Studiengangs Geodäsie und Geoinformatik
2013,2

Eine Übersicht aller bisher in dieser Schriftenreihe erschienenen Bände finden Sie am Ende des Buches.

Orbital Effects in Spaceborne Synthetic Aperture Radar Interferometry

by

Hermann Bähr

Dissertation, Karlsruher Institut für Technologie
Fakultät für Bauingenieur-, Geo- und Umweltwissenschaften, 2013
Referenten: Prof. Dr.-Ing. Dr.-Ing. E. h. Günter Schmitt
             Prof. dr. ir. Ramon F. Hanssen
             Prof. Dr.-Ing. Dr. h. c. Bernhard Heck

Diese Arbeit ist gleichzeitig veröffentlicht in:
Deutsche Geodätische Kommission, Reihe C, Band 719
Verlag der Bayerischen Akademie der Wissenschaften, München, 2013
ISBN 978-3-7696-5131-7

Impressum

Scientific
Publishing

Karlsruher Institut für Technologie (KIT)
KIT Scientific Publishing
Straße am Forum 2
D-76131 Karlsruhe

KIT Scientific Publishing is a registered trademark of Karlsruhe
Institute of Technology. Reprint using the book cover is not allowed.

www.ksp.kit.edu

Print on Demand 2013

ISSN 1869-9669
ISBN 978-3-7315-0134-3

Orbital Effects
in Spaceborne Synthetic Aperture Radar Interferometry

Zur Erlangung des akademischen Grades eines

DOKTOR-INGENIEURS

von der Fakultät für

Bauingenieur-, Geo- und Umweltwissenschaften

des Karlsruher Instituts für Technologie

genehmigte

DISSERTATION

von

Dipl.-Ing. Hermann Bähr

aus Bad Hersfeld

Tag der mündlichen Prüfung: 22. Januar 2013

Referent: Prof. Dr.-Ing. Dr.-Ing. E. h. Günter Schmitt
Korreferent: Prof. dr. ir. Ramon F. Hanssen
Korreferent: Prof. Dr.-Ing. Dr. h. c. Bernhard Heck

Karlsruhe 2013

Summary

Orbit errors usually form a minor contribution to the error budget of spaceborne repeat-pass synthetic aperture radar interferometry (InSAR). However, inaccurately determined satellite trajectories can occasionally have a very significant effect on interferometric products and distort the large-scale component of the deformation signal. It is thus indispensable to be aware of the underlying mechanisms when applying InSAR to deformation monitoring and to eventually consider dedicated corrections. Against this background, the impact of orbit errors on InSAR processing is comprehensively analysed.

Following a brief introduction to InSAR processing, a both quantitative and qualitative characterisation of expectable orbit errors is provided. The accuracy of available orbit products is reviewed and evaluated by gathering global quality indicators originating from validation campaigns. This survey is complemented by a parametric characterisation of the interrelation between relative orbit errors or baseline errors, respectively, on the one hand, and error signals in the interferometric phase or coregistration offsets, respectively, on the other hand.

Based thereupon, approaches to reversely infer baseline corrections from residual phase patterns are reviewed and evaluated with particular attention to the approximation quality of different parameterisations. As a result, two estimators with optimised properties are described in detail: a least squares estimator requiring prior unwrapping and a gridsearch estimator that can handle the wrapped phase. Both are based on the same functional model, accounting for baseline errors by two parameters: the error component perpendicular to the line of sight and the error in the rate of change of the parallel component.

The methodology is generalised by adjusting baseline error estimates in an overdetermined network of linearly dependent interferometric combinations of images. Thus, systematic biases, for instance due to unwrapping errors, can be detected and iteratively eliminated. Regularising the solution by a minimum-norm condition also enables the inference of quasi-absolute orbit errors that refer to individual acquisitions. Testing this approach on a sample Envisat data set involves the evaluation of different stochastic models and concepts of hierarchical organisation. Whereas the least squares estimator produces a consistent solution, gridsearch estimates turn out to be unreliable in specific cases.

The study of orbit error correction approaches is concluded by an outlook on potential application scenarios. It is further complemented by analysing some related error mechanisms that likewise stem from inaccurate modelling of the acquisition geometry. Thus, the effects of timing errors and clock errors are characterised, and the significance of decorrelation due to orbit convergence is investigated. A whole chapter is dedicated to the effect of unmodelled reference frame motion on InSAR deformation estimates. The resulting bias is predicted for Envisat acquisitions at various locations on the globe, and three correction approaches are proposed.

Ausführliche Zusammenfassung

Interferometrie von Radaraufnahmen mit synthetischer Apertur (InSAR) ist ein mittlerweile etabliertes Verfahren zur flächenhaften Erfassung von Bodendeformationen. Eine wesentliche Herausforderung bei der InSAR-Prozessierung besteht darin, sämtliche Signalkomponenten, die das Deformationssignal in den Interferogrammen überlagern und somit verschleiern, bestmöglich zu modellieren. Die vorliegende Arbeit befasst sich ausführlich mit jenen Signalanteilen, die aus einer ungenauen oder fehlerhaften Rekonstruktion der Aufnahmegeometrie bei der interferometrischen Prozessierung resultieren. Der Schwerpunkt liegt auf der Untersuchung von Restfehlern bei der präzisen Bestimmung der Satellitenflugbahn, die im Interferogramm ein nahezu lineares Fehlersignal hervorrufen. Neben einer Charakterisierung der Wirkmechanismen wird eine Abschätzung der Relevanz vorgenommen, und mögliche Korrekturansätze werden aufgezeigt.

Während alle Betrachtungen möglichst allgemein gehalten sind, werden aufgrund der Vielzahl verfügbarer Sensoren und Anwendungsszenarien einige Einschränkungen vorgenommen. Berücksichtigt wird demnach ausschließlich die interferometrische Auswertung von SAR-Aufnahmen satellitengetragener Sensoren zur Erfassung von Deformationen der Erdoberfläche. Auf die Anwendbarkeit der InSAR-Technik zur Generierung digitaler Oberflächenmodelle wird nicht gesondert eingegangen, was etwa mit der zunehmenden Verfügbarkeit flächendeckender Höhenmodelle gerechtfertigt wird. Beispielrechnungen verwenden Parameter der Envisat-Mission und haben damit auch für die ERS-Satelliten hinreichende Gültigkeit. Um ein möglichst breites Spektrum aktueller und künftiger SAR-Sensoren abzudecken, wird zusätzlich auf Besonderheiten der Missionen Radarsat-1/2, ALOS, TerraSAR-X und Sentinel-1 eingegangen.

Grundlegende Kenntnisse in den Bereichen InSAR, Ausgleichungsrechnung und Signalverarbeitung voraussetzend, beginnt die Abhandlung mit einem konzisen Überblick über das InSAR-Messprinzip. Einzelne Verarbeitungsschritte zur Bildung von Interferogrammen werden am Beispiel des Delfter Objektorientierten InSAR-Prozessors (DORIS) erläutert. Aufgrund einer Zerlegung der interferometrischen Phase in Bestandteile bezüglich Geometrie, Deformation, Atmosphäre und Messrauschen werden die stochastischen Eigenschaften einzelner Komponenten im Hinblick auf deren Trennbarkeit diskutiert. Von besonderem Interesse ist die Unterscheidung zwischen Deformationssignal und Störsignalen aufgrund fehlerhaft rekonstruierter Satellitenbahnen. Eine klare Trennung kann nur im Rahmen einer Zeitreihenanalyse erfolgen, für die es zwei etablierte Ansätze gibt: die alleinige Auswertung zeitlich persistenter Punktstreuer (PS-InSAR) sowie die Beschränkung auf Interferogramme mit kurzen Basislinien. Beide Varianten werden ausführlich vorgestellt. Den Abschluss des Grundlagenkapitels bildet eine kurze Übersicht über konkrete Szenarien, in denen eine fehlerhaft angenommene Satellitenbahn das Ergebnis einer InSAR-Deformationsanalyse empfindlich beeinflussen kann.

Um ein vertieftes Verständnis für die Bedeutung von Satelliten-Bahnfehlern für die InSAR-Prozessierung zu vermitteln, erfolgt eine ausführliche Charakterisierung der entsprechenden Wirkmechanismen. Dazu wird zunächst die Methodik der präzisen Bahnbestimmung erläutert, und es werden Untersuchungsergebnisse zu erreichten Genauigkeiten für die einzelnen SAR-Missionen zusammengestellt. Daraus zeigt sich, dass die Qualität der Bahndaten sehr heterogen ist und sich deren Genauigkeiten im Bereich von wenigen Zentimetern (TerraSAR-X) bis zu einigen Metern (Radarsat-1) bewegen können. Derartige Angaben sind jedoch stets kritisch zu beurteilen, da erfolgte Validierungen zumeist nicht vollständig unabhängige Messdaten verwenden, verfügbare Qualitätsmaße nicht einheitlich definiert sind und die räumlich sowie zeitlich variable Bahngenauigkeit mit nur einer einzigen Kenngröße nicht adäquat charakterisiert werden kann.

Für die InSAR-Technik sind weniger absolute sondern ausschließlich relative Fehler in den Trajektorien beider Überflüge von Bedeutung. Diese können auch als Fehler der dreidimensionalen Basislinie aufgefasst werden, die die jeweiligen Aufnahmezentren miteinander verbindet. Insbesondere können Basislinienfehler senkrecht zur Blickrichtung des Sensors mit Interferenzstreifen parallel zur Flugrichtung assoziiert werden. Ähnlich verursachen Fehler in der Änderungsrate der Basislinienkomponente in Blickrichtung Interferenzstreifen senkrecht zur Flugrichtung. Bei flachem Gelände lassen sich die durch Bahnfehler verursachten Störsignale im Interferogramm in sehr guter Näherung als Überlagerung dieser beiden Komponenten charakterisieren.

Zusätzlich machen sich Basislinienfehler auch bei der Koregistrierung bemerkbar, und zwar dahingehend, dass die aus der Aufnahmegeometrie abgeleitete Zuordnung korrespondierender Bildkoordinaten nicht mit den aus einer Kreuzkorrelation der Amplitudenbilder ermittelten Ablagen übereinstimmt. Verglichen mit den Störsignalen in der interferometrischen Phase ist dieser Effekt jedoch wesentlich schwächer ausgeprägt und bleibt daher für alle weiteren Betrachtungen unberücksichtigt.

Neben Basislinienfehlern werden auch die Auswirkungen von Zeitgebungsfehlern, Frequenzfehlern und konvergenten Trajektorien untersucht. Dabei wird festgestellt, dass etwa grobe Fehler in der den Bilddaten annotierten Signallaufzeit oder Abweichungen der Radar-Trägerfrequenz von ihrem Nominalwert nahezu lineare Phasenartefakte verursachen. Beide Effekte können in Ausnahmefällen, sofern sie unentdeckt bleiben, das Interferogramm und daraus abgeleitete Parameter signifikant verfälschen. Im Zuge einer spezifischen Betrachtung der Envisat-Flugbahn wird zudem exemplarisch nachgewiesen, dass die geometrische Dekorrelation der Interferogramme aufgrund etwaiger Konvergenz der Trajektorien zweier Überflüge im allgemeinen vernachlässigbar ist.

Im Anschluss an diese eingehenden Betrachtungen beschäftigt sich die Arbeit im Kern mit der parametrischen Schätzung von Bahnfehlern aus residuellen Störsignalen in Interferogrammen. Dazu werden zunächst existierende Ansätze hinsichtlich wesentlicher Charakteristika vergleichend evaluiert. Von besonderem Interesse sind die Approximationsgüte verschiedener Parametrisierungen sowie die Auswahl geeigneter Interferogrammpixel als Beobachtungen bzw. Datengrundlage der Schätzung. Ein wichtiges Kriterium ist dabei die Resistenz der Schätzwerte gegenüber Ausreißern, die besonders von der räumlichen Verteilung der Beobachtungen bzw. Beobachtungsgewichte abhängt und durch Überparametrisierung negativ beeinflusst werden kann.

Zwei Schätzverfahren mit optimierten Eigenschaften werden im Detail beschrieben. Beide verwenden als Beobachtungen die interferometrischen Phasen einer Auswahl gleichmäßig verteilter, kohärenter Pixel und beschreiben das Fehlersignal durch zwei Parameter: den Basislinienfehler senkrecht zur Blickrichtung des Radars und die Änderungsrate des Fehlers in Blickrichtung. Einer der Schätzer minimiert die Quadratsumme der Phasenresiduen, was eine vorherige Mehrdeutigkeitslösung durch Phasenabwicklung erfordert. Der andere sucht im Parameterraum nach derjenigen Lösung, der die komplexe Summe der Phasenresiduen maximiert. Dieser sogenannte Suchgitter-Ansatz erfordert keine Phasenabwicklung, ermöglicht aber im Gegensatz zur Schätzung nach kleinsten Quadraten keine differenzierte stochastische Modellbildung.

Zur Steigerung der Resistenz gegenüber Ausreißern in einzelnen Interferogrammpixeln wird für den Kleinste-Quadrate-Schätzer iteratives Data-Snooping vorgeschlagen. Eine ungleich wichtigere Möglichkeit zur Qualitätskontrolle bietet die Überbestimmung der Bahnfehler einzelner Aufnahmen in einem Netz linear abhängiger Interferogrammkombinationen. Darin können Widersprüche relativer Basislinienfehler getilgt und mit geeigneter Datumsverfügung absolute Bahnfehler abgeleitet werden. Zur Erhöhung der Zuverlässigkeit wird die Konsistenz der Beiträge einzelner Interferogramme statistisch getestet, und etwaige, beispielsweise durch fehlerhafte Phasenabwicklung entstandene Ausreißer können detektiert und sukzessive eliminiert werden.

Die beschriebenen Verfahren werden an einem Datensatz aus 31 Envisat-SAR-Bildern getestet. Die verwendeten 163 Interferogrammkombinationen werden so ausgewählt, dass großräumige Phasensprünge in der abgewickelten Phase aufgrund ausreichend hoher Kohärenz ausgeschlossen werden können. Während die Ausgleichung der Basislinienfehler nach kleinsten Quadraten eine durchweg konsistente Schätzung liefert, erweist sich der Suchgitter-Ansatz empfindlich gegenüber überlagernden atmosphärischen Störsignalen, und es ergeben sich widersprüchliche Fehlerparameter für einzelne Interferogramme. Zwar ist es möglich, diese Widersprüche durch Data-Snooping sukzessive zu eliminieren, jedoch konvergiert die Lösung nicht gegen die Schätzung nach kleinsten Quadraten. Für diese wird hingegen durch Simulation von Phasensprüngen in einzelnen Interferogrammen nachgewiesen, dass Ausreißer verlässlich detektiert werden können.

Eine besondere Herausforderung stellt die Wahl eines geeigneten stochastischen Modells bei der Ausgleichung nach kleinsten Quadraten dar. Einer möglichst exakten Modellbildung stehen hier die Restriktionen gegenüber, dass sich einerseits eine inhomogene Gewichtung negativ auf die Robustheit des Schätzers auswirkt und dass andererseits eine konsistente Schätzung von Kovarianzen bei räumlich-langperiodischen atmosphärischen Signalen nicht möglich ist. Auch eine strenge Berücksichtigung algebraischer Korrelationen ist nicht ohne weiteres umsetzbar. Es werden daher unter Vernachlässigung der Korrelation zwischen Interferogrammen drei Kompromisslösungen evaluiert: ein Ansatz mit vollständig unkorrelierten Beobachtungen und zwei individuell angepasste Modelle für Kovarianzfunktionen zur Beschreibung räumlich-isotroper Korrelationen. Obwohl die statistische Validierung der Modelle auf Beobachtungsebene in keinem der drei Fälle gelingt, kann zumindest die Gültigkeit des Ausreißertests auf Interferogrammebene nachgewiesen werden.

Die Untersuchungen beinhalten auch den Vergleich zweier Varianten der hierarchischen Organisation. Einerseits kann die Ausgleichung in geschlossener Form erfolgen, so dass absolute Bahnfehler für jede SAR-Aufnahme direkt aus den einzelnen Pixeln sämtlicher Interferogramme geschätzt werden. Es ist aber auch ein zweistufiger Ansatz möglich, bei dem interferogrammweise Basislinienfehler als Zwischenergebnisse auftreten. Trotz geringfügiger Abweichungen in funktionaler und stochastischer Modellbildung unterscheiden sich die Ergebnisse nicht wesentlich, so dass beide Ansätze als äquivalent angesehen werden können.

Abschließend wird begründet, wieso die vorgestellten Verfahren zur Schätzung von Bahnfehlern einen Beitrag zur InSAR-Prozessierung liefern können. Obwohl die Bahnen neuerer SAR-Satelliten mit ausreichender Genauigkeit bestimmt werden können, wird es auch in naher Zukunft noch Bedarf an der Prozessierung von Daten älterer Satelliten oder von historischen Aufnahmen geben. Zudem kann eine gleichbleibend gute Qualität der Bahnbestimmung nicht garantiert werden, und insbesondere für Analysen in Nahezu-Echtzeit wird die erforderliche Genauigkeit nicht erreicht. Für Anwendungen mit höchsten Genauigkeitsanforderungen werden Bahnfehler auch in Zukunft nicht vernachlässigbar sein, und entsprechende Schätzverfahren können einen wichtigen Beitrag zu einer integrierten Modellbildung liefern.

Ein letztes Kapitel der Arbeit ist der Auswirkung der kontinuierlichen tektonischen Plattenbewegung auf die InSAR-Prozessierung gewidmet, die im allgemeinen vernachlässigt wird. So wird aufgrund der Relativbewegung der Erdoberfläche bezüglich des Koordinatenrahmens, in dem die Bahndaten gegeben sind, das Interferogramm auf ähnliche Weise verfälscht wie durch Basislinienfehler. Das resultierende Fehlersignal ist nahezu linear und aufgrund der Gleichförmigkeit der Plattenbewegung proportional zum zeitlichen Abstand der beiden SAR-Aufnahmen. Unter der Annahme, dass die Bahndaten im *Internationalen Terrestrischen Referenzrahmen* (ITRF) gegeben sind, wird die Verfälschung der Deformationsmessung für 840 global verteilte ITRF-Stationen prädiziert. Die resultierenden Fehlerprognosen sind regional verschieden und können für großräumige Deformationsanalysen mit hohen Genauigkeitsanforderungen hochrelevant sein.

Zur Korrektur dieses sogenannten *Referenzrahmeneffektes* werden drei Varianten vorgeschlagen, die alle auf einer Datumstransformation der Bahndaten beruhen. Die einfachste Möglichkeit besteht in einer simplen Parallelverschiebung der Flugtrajektorien in Abhängigkeit vom Aufnahmezeitpunkt. Andere Ansätze bestehen in der Drehung des Bahnkoordinatensystems um einen Eulerpol bzw. in einer allgemeinen Ähnlichkeitstransformation mit allen sechs hier relevanten Freiheitsgraden. Die Unterschiede der drei Korrektionen hinsichtlich ihrer Approximationsgüte werden ausführlich diskutiert.

Contents

1. Introduction

Spaceborne repeat-pass synthetic aperture radar interferometry (InSAR) is a promising technique for measuring deformation of the earth's surface. It stands out due to its dense spatio-temporal coverage and the related cost-efficiency. Advancing its potential involves getting hold of all kinds of error sources and associated signals that may obscure the inferable surface deformation. Inaccuracies in satellite orbit data are one of the essential contributions to the error budget. They induce an error signal that is almost linear in space and would suggest a tilt of the surface when interpreted as ground deformation. Accordingly, neglecting orbit errors becomes critical if the spatial extent of the area of interest is large.

Orbit errors constitute by far not the most relevant error source in SAR interferometry, and their relevance is continuously decreasing due to recent advances in precise orbit determination. For this reason they have rarely been in the focus of research and are mostly covered as one among multiple challenges of InSAR processing. Whereas it is generally appropriate to consider the signal of interest together with all significant error signals, such a comprehensive approach cannot give attention to particular aspects in depth. In order to fill this gap, this thesis explicitly focusses on orbit errors and orbit-related effects, reviewing existing contributions and complementing previous achievements.

All considerations in the following chapters are limited to spaceborne repeat-pass InSAR and its application to deformation monitoring. The effect of orbit errors on InSAR-generated digital elevation models (DEM) is not explicitly addressed, since it is becoming less relevant in view of the increasing availability of high-quality DEM products from single-pass acquisitions with an outstanding relative orbit accuracy. To account for the whole variety of sensors and acquisition modes, analytical considerations are kept as generic as possible. Sample computations are specialised on the *Environmental Satellite* (Envisat) and are thus practically conferrable to the *European Remote Sensing Satellites* (ERS-1 and ERS-2).

It is not intended to provide a general introduction to the InSAR technique, for which dedicated textbooks and topical reviews are recommended (HANSSEN, 2001; BAMLER AND HARTL, 1998; MASSONNET AND FEIGL, 1998; ROSEN ET AL., 2000; ZEBKER ET AL., 2000; XIA, 2010; RICHARDS, 2009, ch. 6). The reader should be familiar with basic principles of SAR and InSAR as well as signal processing and adjustment theory. Thus, chapter 2 provides only a brief review on the fundamentals of state-of-the-art InSAR processing with particular emphasis on orbit-related aspects in order to introduce some terminology and revisit relevant methodology.

In chapter 3 the interrelations of orbit errors and corresponding error signals in interferometric products are characterised by forward modelling from a biased satellite trajectory. Besides baseline errors and orbit convergence, also errors in timing and frequency are covered, and their effect on both the interferometric phase and coregistration offsets is investigated. Finally, the significance of differently parameterised types of error is evaluated with the objective to assess the potential of dedicated correction approaches.

The core of the thesis is formed by chapters 4 through 6, which are concerned with the inverse problem of inferring orbit errors from residual interferometric phase patterns. Existing approaches are reviewed and evaluated in various respects, and two optimised estimators are presented in detail. Particular emphasis is placed on the network approach, which provides a framework to reliably identify outliers by a joint estimation from redundant interferometric combinations. Conclusions, outlook and recommendations regarding this complex of three chapters are placed at the end of chapter 6. (Parts of chapters 4 through 6 overlap with previous publications: BÄHR AND HANSSEN, 2010, 2012.)

Chapter 7 addresses a side topic that emerged from the research on orbit errors: The neglect of relative motion of the orbit reference frame in InSAR processing can induce a significant error signal into the interferometric measurement. The underlying mechanism is characterised, and three different correction approaches are proposed. (Chapter 7 partly overlaps with one previous publication: BÄHR ET AL., 2012.)

2. Deformation Monitoring with Spaceborne InSAR

The intention of this chapter is neither to provide a comprehensive introduction to InSAR nor to be a complete review of hitherto developed methods. Rather a brief overview is given, introducing notation and concepts. Emphasis is placed on aspects that are meaningful for the subsequent analysis of the effect of orbit inaccuracies on detection and mapping of surface deformation. Considerations are restricted to the application of spaceborne repeat-pass InSAR for detection and mapping of large-scale deformation phenomena, starting from zero-Doppler focussed *Single Look Complex* (SLC) imagery.

Section 2.1 addresses the basics of SAR processing and focussing, which are explained in its entire scope in dedicated textbooks (CURLANDER AND MCDONOUGH, 1991; CUMMING AND WONG, 2005) and the habilitation treatise of MOREIRA (2000, in German). Interferometric concepts are outlined in section 2.2, and section 2.3 gives some details on processing. In section 2.4, time series approaches are addressed, focussing on their capability to handle orbital errors. The chapter concludes with a summary of aspects why orbit errors can be critical for deformation monitoring.

2.1. SAR Measurement Principle

Spaceborne SAR acquisitions are taken by a radar instrument (sensor) that is installed on a satellite (platform) orbiting the earth on a smooth trajectory. In equidistant intervals, a side-looking antenna transmits linearly frequency modulated radar pulses (chirps) towards the surface. Their backscattered echoes are generally received by the very same antenna, subsequently quadrature demodulated and digitised. The result are *raw data*: an image matrix of complex numbers $z = Ae^{i\psi}$ with annotated times of transmission and reception. Raw data are not conveniently interpretable, because every target on the ground is illuminated by several subsequent pulses.

Between transmission and reception of a pulse, the platform displaces by some tens of metres with respect to a target on the ground. This relative motion causes a Doppler-like effect. The signal response of the target is shifted by a Doppler frequency, which is an equivalent measure for the squint angle β under which it is illuminated (see figure 2.1a). The Doppler centroid frequency f_{DC} characterises target responses from the centre of the radar beam. At the point of closest approach ($\beta = 0$) the Doppler frequency is zero. Usually, the beam squint is yaw-steered in a way that $f_{DC} \approx 0$, also taking into account the earth rotation. Residual deviations can be estimated from the data (BAMLER AND SCHÄTTLER, 1993, p. 91).

To enable further handling and interpretation, raw data are *focussed* by image processing techniques. The result is a complex image matrix, the pixels of which can be mapped one-to-one to a generally rectangular resolution cell on the ground. In the SLC format, which is a common standard for focussed SAR data, the data are sampled in a zero-Doppler-azimuth/slant-range coordinate system (GEUDTNER, 1995; see also figure 2.1b). The *azimuth* coordinate specifies the zero-Doppler plane of the resolution

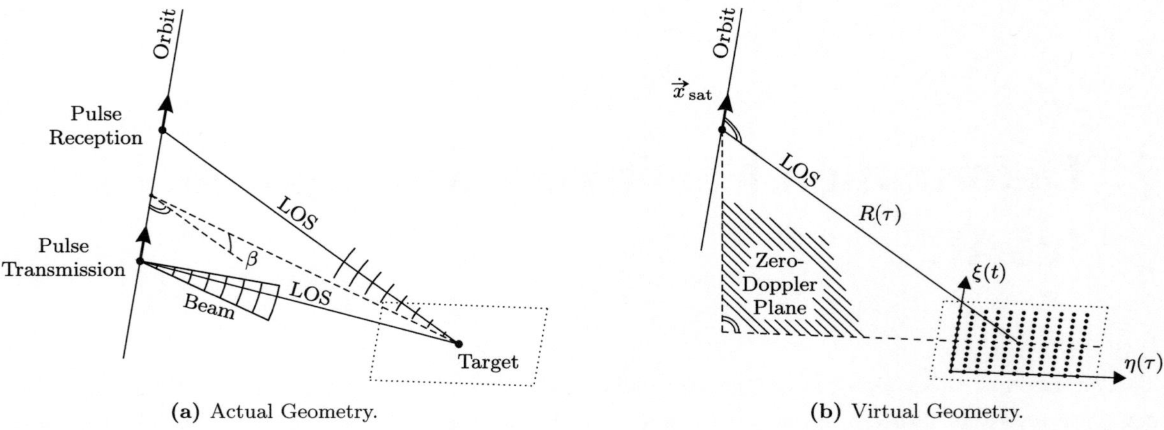

(a) Actual Geometry.   (b) Virtual Geometry.

Figure 2.1.: SAR Acquisition Geometry. **(a)** The platform displaces between transmission and reception of a pulse. The squint angle β is defined in a plane defined by the line of sight (LOS) to the target and back, measured halfway between transmission and reception. **(b)** The platform maintains its positions between transmission and reception of the pulse (start-stop approximation).

cell, i. e., a plane that is perpendicular to the orbit trajectory. The *range* coordinate is a measure of the distance to the orbit within this plane.

When dealing with SLC data, the following virtual acquisition geometry may conveniently be assumed (start-stop approximation, BAMLER AND SCHÄTTLER, 1993): The sensor transmits a modulated pulse perpendicular to its orbit and maintains its position until the reception of the echo. Then it moves on a few metres along the orbit to transmit the subsequent pulse. Thus, every ground resolution cell is imaged only once and can be identified by the virtual transmission time t of the associated pulse (azimuth time, slow time) and the virtual two-way signal travel time τ (range time, fast time). With the speed of light c, τ can be converted to range:

$$R = \frac{c}{2}\tau \ . \tag{2.1}$$

Moreover, pixel coordinates (ξ, η) can be inferred from the timing (t, τ):

$$
\begin{aligned}
\xi(t) &= (t - t_1) \cdot f_{\text{PRF}} &&\text{(azimuth)}\\
\eta(\tau) &= (\tau - \tau_1) \cdot f_{\text{RSR}} &&\text{(range)} \ .
\end{aligned}
\tag{2.2}
$$

t_1 is the acquisition start time, τ_1 the two-way signal travel time from the orbit to the first sampled pixel (sampling window start time, SWST), f_{PRF} the pulse repetition frequency (PRF), and f_{RSR} the range sampling rate (RSR).

The described acquisition principle is specific to the stripmap mode, which is the standard mode for most sensors. However, the start-stop approximation can be used to reproduce the target locations for any SLC data set, regardless of the acquisition mode.

2.2. Interferometry

Detecting surface displacements by interferometry requires two SAR images acquired at different times T. Having introduced azimuth time t and range time τ in the previous section, T is a third timescale to be used within the scope of this thesis. Whereas t and τ are rather substitutes for spatial coordinates in azimuth and range, T refers to the acquisition time of a whole image in the long term context. The temporal separation of two acquisitions is called the temporal baseline B_T. Within the scope of this thesis, temporal derivatives of a variable X will be denoted by:

$$\dot{X} := \frac{\partial X}{\partial t} \qquad \text{and} \qquad X' := \frac{\partial X}{\partial T} \, , \tag{2.3}$$

respectively.

As the sampling grids of two SAR images are generally not congruent, interferogram formation requires prior coregistration and resampling of one image to the geometry of the other. The latter image, which defines the reference geometry, is termed *master* (M), and the resampled image is referred to as *slave* (S). The complex interferogram z_I is then computed by multiplying the master image by the complex conjugate of the slave image:

$$z_\mathrm{I} = z_\mathrm{M} z_\mathrm{S}^* = A_\mathrm{M} A_\mathrm{S} e^{i(\psi_\mathrm{M} - \psi_\mathrm{S})} = A_\mathrm{I} e^{i\phi} \, , \tag{2.4}$$

where $(\cdot)^*$ denotes the complex conjugate. Thus, the interferometric phase ϕ is obtained. In order to analyse different signal components, it is convenient to decompose the phase into four contributions (FERRETTI ET AL., 2000):

$$\phi = \phi_\mathrm{geom} + \phi_\mathrm{defo} + \phi_\mathrm{atmo} + \phi_\mathrm{noise} \, . \tag{2.5}$$

The geometric phase ϕ_geom represents the contribution of the acquisition geometry at the time of the master acquisition. The component ϕ_defo accounts for target displacements in direction of the line of sight between the two acquisitions. The effect of atmospheric propagation delay is considered by ϕ_atmo. ϕ_noise subsumes all remaining contributions.

In the following, the components will be discussed in particular.

2.2.1. Geometric Phase

The geometric term ϕ_geom makes by far the largest contribution to the interferometric phase. It is determined by the spatial positions of master (M), slave (S) and the target (P) and can be characterised by the interferometric baseline B (see figure 2.2), which is decomposable into the components $B_\parallel$ in ranging direction and $B_\perp$ perpendicular to the line of sight (LOS). In order to reveal the information content of an interferogram, it is common practice to eliminate the contribution of geometry. This is achieved by subtracting a reference phase (HANSSEN, 2001, p. 116):

$$\phi_\mathrm{ref} = -\frac{4\pi}{\lambda} \left(R_\mathrm{M,ref} - R_\mathrm{S,ref} \right) . \tag{2.6}$$

$R_\mathrm{M,ref}$ and $R_\mathrm{S,ref}$ are the ranges to a reference surface, the selection of which depends on the specific application, and λ is the carrier wavelength. If the objective is the measurement of ground displacements, this reference surface is an approximation of the terrain surface, mostly parameterised by a digital elevation model (DEM). Subtracting the reference phase from the interferogram eliminates the geometric phase completely, except for residual errors $\delta\phi_\mathrm{topo}$ and $\delta\phi_\mathrm{orb}$:

$$\varphi := \phi - \phi_\mathrm{ref} = \delta\phi_\mathrm{topo} + \delta\phi_\mathrm{orb} + \phi_\mathrm{defo} + \phi_\mathrm{atmo} + \phi_\mathrm{noise} \, . \tag{2.7}$$

As the contribution of inaccuracies or approximation errors $\delta\phi_\mathrm{topo}$ of the DEM is identical for every acquisition, it can be easily estimated and eliminated in time series approaches. This is not the case for orbit errors $\delta\phi_\mathrm{orb}$, which induce almost linear ramps into an interferogram, behaving randomly for individual acquisitions. Their characteristics will be analysed in detail in chapter 3.

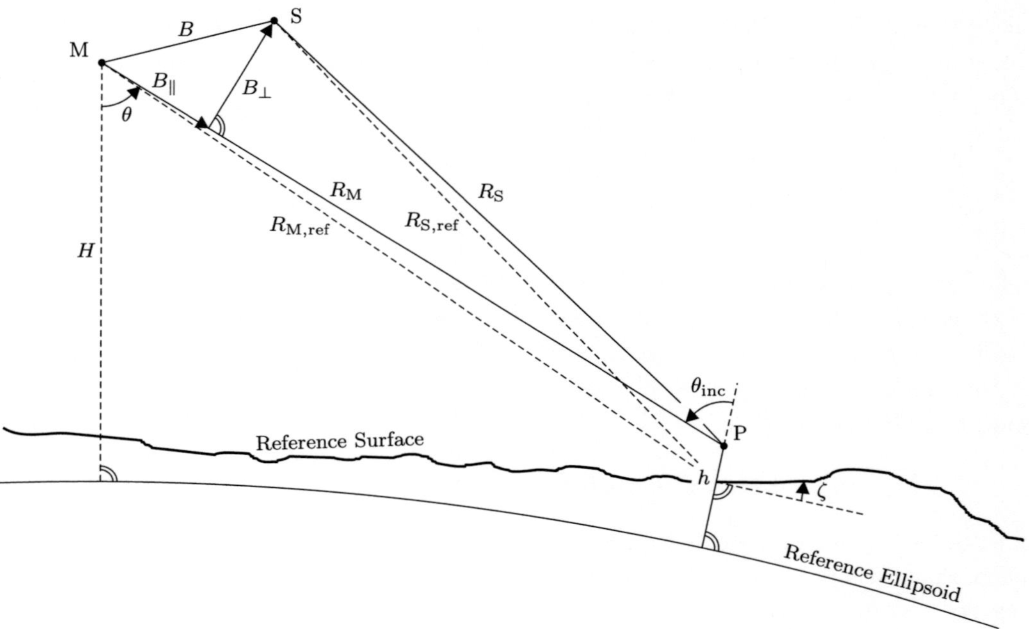

Figure 2.2.: InSAR Acquisition geometry of spaceborne across-track interferometry.

Identifying the terrain surface as reference surface is not a common approach, but it is appropriate if deformation is the signal of interest and a sufficiently accurate DEM is available. It is rather common practice to subdivide the geometric phase into two components: a contribution of the reference ellipsoid (flat earth phase) and a contribution of the topography above the ellipsoid (topographic phase). If the objective of interferometry is topographic mapping, the topographic phase is isolated by identifying the ellipsoid as reference surface. Thus, an elevation model can be inferred from the topographic phase by exploiting its sensitivity to the elevation h (HANSSEN, 2001, p. 37):

$$\frac{\partial \phi_{\mathrm{geom}}}{\partial h} = -\frac{4\pi}{\lambda}\frac{\Delta\theta}{\sin\theta_{\mathrm{inc}}} = -\frac{4\pi}{\lambda}\frac{B_\perp}{R\sin\theta_{\mathrm{inc}}} \,. \tag{2.8}$$

$\Delta\theta$ is the look angle difference, θ_{inc} is the local ellipsoidal incidence angle and $R := R_M$ (see figure 2.2). But as topographic mapping is not of primary relevance to the present thesis, this application will not be considered any further.

Interferograms for deformation mapping are often referred to as *differential interferograms*, and the associated processing technique as *differential InSAR* (DInSAR). These names were created when the availability of global DEM with a suitable resolution was poor and the reference phase could only be computed with respect to the ellipsoid. Surface displacements could only be retrieved if the topographic contribution was derived from an additional, complementary interferogram (three- or four-pass method, respectively; ZEBKER ET AL., 1997). Applications for which no appropriate DEM is available occur significantly less frequently since the release of DEM products of the *Shuttle Radar Topography Mission* (SRTM) or generated from data of the *Advanced Spaceborne Thermal Emission and Reflection Radiometer* (ASTER). They are expected to become rare as soon as elevation data from the *TanDEM-X* mission (KRIEGER ET AL., 2007) will be released. But nevertheless, also interferograms computed with respect to a DEM-defined reference surface are called ”differential”, because they can be considered as the difference of a measured interferogram ϕ and a synthetic interferogram ϕ_{ref} simulated from orbit data and a DEM. Strictly speaking, this reasoning can be applied to any interferogram, from which the reference phase has been subtracted, regardless if the reference surface is defined by a DEM or merely the bare ellipsoid. Hence, the notion of ”differential” interferograms seems obsolete and will not be used within the scope of this thesis.

2.2.2. Deformation

Ground deformation can be measured as the relative displacement of scatterers in the direction of the line of sight within the extent of an interferogram. If the master acquisition predates the slave acquisition, a displacement D towards the sensor implies a decrease in range and thus according to eq. (2.6) a decrease in phase. Hence, the phase contribution of deformation reads:

$$\phi_{\mathrm{defo}} = -\frac{4\pi}{\lambda} D \ . \tag{2.9}$$

The sign inversion between D and ϕ allows to conveniently associate a positive displacement with ground uplift and a negative displacement with subsidence. This convention is not consistently made in literature though, where definitions with positive sign (FERRETTI ET AL., 2001; HANSSEN, 2001; HOOPER ET AL., 2007) and negative sign (KAMPES, 2006; FEIGL AND THURBER, 2009; KETELAAR, 2009; HOOPER ET AL., 2010) can be found.

Deformation signals ϕ_{defo} can be observed at almost all temporal and spatial scales. Coseismic displacement occurs instantly, whereas interseismic creep can cover a very long timespan of years or even decades. Landslides and soil subsidence typically induce changes with an intermediary temporal behaviour. As to the spatial extent, most deformation signals affect areas in the order of hundreds of metres up to tens of kilometres. Some particular effects like the accumulation of tectonic strain can affect very large regions of hundreds of kilometres.

2.2.3. Atmosphere

Atmospheric signals ϕ_{atmo} can be subdivided into contributions of the ionosphere and the subjacent, electrically uncharged neutrosphere. The neutrospheric part is clearly the dominant one and depends on the individual weather conditions at the epochs of master and slave acquisition. Differences in propagation delay of several centimetres can result from refractivity variations. These are by far most pronounced in the troposphere, which covers the lower part of the neutrosphere. Hence, the neutrospheric contribution is commonly referred to as tropospheric contribution.

HANSSEN (2001, p. 131) distinguishes two types of neutrospheric signals: turbulent mixing and vertical stratification. Turbulent mixing is driven by turbulent processes at spatial scales above 500 m (HANSSEN, 2001, p. 143). Inhomogeneous distributions of water vapour (wet component) dominate refractivity variations at short spatial wavelengths in the order of some kilometres. Lateral gradients of pressure or temperature (hydrostatic component) are smaller and generally occur at larger scales that may even exceed the size of an interferogram (DING ET AL., 2008, p. 5430).

The stratigraphic signal is the outcome of differing vertical refractivity profiles of master and slave, which affect the phase in case of significant variations of terrain height. Consequently, the associated relative propagation delay is correlated with height and can be in the order of 1-2 cm delay per kilometre height difference (HANSSEN, 2001, p. 152; CAVALIÉ ET AL., 2007; ELLIOT ET AL., 2008). Both effects – turbulence and stratification – may be considered temporally uncorrelated for SAR acquisition intervals exceeding one day (HANSSEN, 2001, p. 131).

Apart from time series approaches, which are addressed in section 2.4, the neutrospheric contribution can be estimated either empirically or by exploiting complementary measurements. Empirical methods subsume geostatistical approaches (e. g., HANSSEN, 2001; KNOSPE AND JÓNSSON, 2010) for the turbulent and regression approaches (e. g., HANSSEN, 2001; CAVALIÉ ET AL., 2007) for the stratigraphic component, but they are not capable to distinguish atmospheric artefacts from other signal contributions with similar statistical properties. This shortcoming does not apply to an integrated processing with complementary measurements from other sensors like terrestrial meteorological instruments, the *Global Positioning System* (GPS), the *Moderate Resolution Imaging Spectroradiometer* (MODIS) or the *Medium Resolution Imaging Spectroradiometer* (MERIS; DING ET AL., 2008). Currently, considerable efforts are made to develop estimation strategies for atmospheric signals based on numerical weather models.

In contrast to neutrospheric effects, very little research has been done on the less significant ionospheric influences. Variations in the concentration of free electrons can cause differential propagation delays that are mapped by the interferometric phase. The integral ionospheric charge is quantified by the *Total Electron Content* (TEC) and measured in *TEC Units* (TECU). The TEC is a globally smooth signal, varying between 0 and 100 TECU (MEYER ET AL., 2006) at spatial wavelengths that are usually larger than 1000 km. TEC magnitudes are distinctly correlated with the sunspot cycle. Especially in equatorial and auroral regions, small scale disturbances can be observed that are usually below 1 TECU and reach down to scales of a few metres (MEYER ET AL., 2006).

Ionospheric effects are most pronounced for L-band InSAR, where a TEC difference of 1 TECU causes a range bias of approximately 32 cm for ALOS-PALSAR (according to HANSSEN, 2001, eq. (6.3.5)). But biases are still significant for C- and X-band, where 1 TECU is equivalent to 15 mm (ERS, Envisat IS2) or 5 mm (TerraSAR-X strip_010), respectively. Typical small-scale ionospheric disturbances (in equatorial and auroral regions) are in the order of 0.3 TECU (MEYER ET AL., 2006), causing a range bias of 4 mm in C-band. These numbers support the conclusion that small scale variations are negligible for C- and X-band in mid-latitudes, where the TEC is relatively undisturbed. This may be different for the long wavelength component, which is additionally supported by the effect of the range-varying signal path length through the ionosphere. The resulting phase trend in range becomes even more significant when combining acquisitions from different stages of the solar cycle (MEYER ET AL., 2006).

2.2.4. Sources of Decorrelation

Besides ϕ_{geom}, ϕ_{defo} and ϕ_{atmo} there are a number of additional effects that can be distinguished in the interferometric phase. However, they can neither be modelled in a deterministic manner, nor do they expose any spatial correlation property. Thus, they are only stochastically relevant for the analysis of orbital effects. The associated signal is mostly perceived as decorrelation and generically subsumed by ϕ_{noise}.

A common measure to assess the correlation of phases from master and slave acquisition is the absolute value $|\gamma|$ of the interferometric coherence (HANSSEN, 2001, p. 96):

$$\gamma = \frac{\mathrm{E}\{z_M z_S^*\}}{\sqrt{\mathrm{E}\{z_M z_M^*\}\,\mathrm{E}\{z_S z_S^*\}}}\,, \quad 0 \leq |\gamma| \leq 1\,. \tag{2.10}$$

Assuming both ergodicity and local spatial stationarity, the actually unknown expectation values $\mathrm{E}\{\cdot\}$ can be approximated for a resolution cell with pixel coordinates (ξ, η) by averaging over a spatial estimation window (HANSSEN, 2001) of odd dimensions (m_ξ, m_η):

$$|\hat{\gamma}|(\xi, \eta) = \frac{\sum_{(i,j)\in W} z_{M,ij} z_{S,ij}^*}{\sqrt{\left(\sum_{(i,j)\in W} z_{M,ij} z_{M,ij}^*\right)\left(\sum_{(i,j)\in W} z_{S,ij} z_{S,ij}^*\right)}}\,, \tag{2.11}$$

where $W = \{(i,j): |\xi - i| \leq (m_\xi - 1)/2 \,\wedge\, |\eta - j| \leq (m_\eta - 1)/2)\}$. ZEBKER AND VILLASENOR (1992) demonstrated that the individual decorrelation effects contribute multiplicatively to the overall coherence:

$$\gamma = \gamma_{rg} \cdot \gamma_{az} \cdot \gamma_{vol} \cdot \gamma_T \cdot \gamma_{system} \cdot \gamma_{processing}\,. \tag{2.12}$$

Based on (HANSSEN, 2001, pp. 98 et seqq.), six effects are distinguished that contribute to ϕ_{noise} and thus to the decorrelation of the interferometric signal: surface decorrelation due to varying viewing directions in range (γ_{rg}) and azimuth (γ_{az}), volume decorrelation (γ_{vol}), temporal decorrelation (γ_T) as well as system-related and processing-induced noise (γ_{system} and $\gamma_{\mathrm{processing}}$, respectively).

The significance of the viewing direction for the signal coherence results from the circumstance that the measured interferometric phase is always a coherent sum of the signal responses of more or less dominant scatterers within one resolution cell. Illuminated from a different angle, the phase contributions of the individual scatterers change, because their ranges to the sensor vary. If the major part of individual scatterers is arranged on a 2D surface, i. e., the terrain surface, the phase change in the coherent sum due to a different viewing direction can be compensated by an according shift of the Radar frequency (GATELLI ET AL., 1994). This means for the illumination by a band limited chirped pulse that the ground reflectivity spectrum is shifted if the incidence angle θ_{inc} changes. Decorrelation (γ_{rg}) increases linearly as the overlap of the spectra of master and slave decreases or $B_\perp$ increases, respectively. RICHARDS (2009, pp. 229 et seqq.) gives a very illustrative description of this effect in the space domain. An analogous effect can be observed in azimuth, where it is attributed to two causes: different squint angles (or Doppler centroids, respectively) and convergent orbit trajectories (HANSSEN, 2001, p. 50). In both cases, decorrelation (γ_{az}) increases with decreasing coincidence of the parts of the ground reflectivity spectrum that are mapped within the processed Doppler bandwidth.

In case of volume scattering, where a large number of scatterers are distributed over a 3D resolution cell, the phase is subject to an almost unpredictable decorrelation (γ_{vol}) when the illumination angle is altered. Mostly, the scattering mechanism is a hybrid form of volume and surface scattering. In contrast to volume decorrelation (γ_{vol}), the amount of surface decorrelation (γ_{rg}, γ_{az}) depends on the convergence of illumination directions between master and slave and is thus predictable from the acquisition geometry under consideration of Doppler centroid estimates. Surface decorrelation can be mitigated by approaches sketched in section 2.3.5.

Temporal decorrelation γ_T is driven by changes in the scattering characteristics of the target of various kinds, e. g., movement of plants in the wind, vegetation growth, meteorological precipitation, freezing/thawing, motion of vehicles or other anthropogenic activities (BAMLER AND HARTL, 1998, p. R43). As some of these effects become stronger with time, the temporal baseline B_T is an indicator for the expectable temporal decorrelation. Based on the stack coherence from (KAMPES, 2006), a rough estimator for surface and temporal decorrelation can support the selection of interferometric combinations:

$$|\hat{\gamma}_{\mathrm{dec}}| = g(B_\perp, B_{\perp,\mathrm{crit}}) \cdot g(\Delta f_{\mathrm{DC}} + \Delta f_{\mathrm{conv}}, B_{\mathrm{az}}) \cdot g(B_T, B_{T,\mathrm{crit}}) , \tag{2.13}$$

where

$$g(x, c) = \begin{cases} 1 - \frac{|x|}{c} & |x| < c \\ 0 & \text{otherwise} \end{cases} . \tag{2.14}$$

$$B_{\perp,\mathrm{crit}} = \frac{\lambda B_{\mathrm{rg}} R_{\mathrm{M}} \tan(\theta_{\mathrm{inc}} - \zeta)}{c} \tag{2.15}$$

(HANSSEN, 2001, p. 102) is the critical baseline at which the interferometric phase of surface scatterers is completely decorrelated. B_{rg} is the chirp bandwidth, ζ the local slope (see figure 2.2), Δf_{DC} the difference in Doppler centroid frequencies, Δf_{conv} the frequency shift due to orbit convergence (see section 2.3.5), and B_{az} is the processed Doppler bandwidth in azimuth (GEUDTNER, 1995, p. 50). The choice of the critical temporal baseline $B_{T,\mathrm{crit}}$ is empirical and depends on the respective region of interest.

The two remaining sources of decorrelation depend neither on geometry nor on the target. System-related decorrelation (γ_{system}) is caused by thermal noise, which is inherent to all electronic measurement

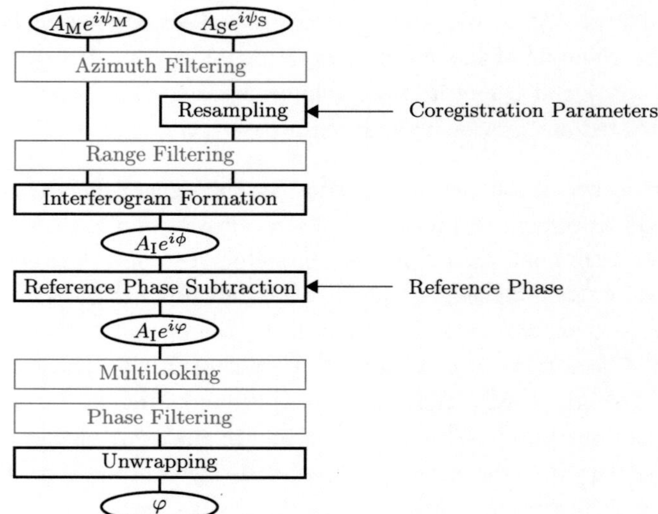

Figure 2.3.: Abstracted processing chain of the DORIS InSAR processor (DUT, 2009). Optional filtering steps are in grey.

devices and imperfections of the hardware. Processing-induced decorrelation ($\gamma_{\text{processing}}$) results from discretisation and interpolation errors or aliasing. It occurs at different stages in the processing chain, including initial sampling of the signal, SAR focussing, slave resampling and filtering.

2.3. Interferometric Processing

After two decades with regularly acquiring spaceborne SAR missions, a number of software packages has been developed for InSAR processing of standardised data products. SIMONETTO (2008) gives an overview of 10 processors, which is not exhaustive and has outdated in the meantime. The *Delft Object-oriented Radar Interferometric Software* (DORIS) has been developed at Delft University of Technology since 1998 (KAMPES ET AL., 2004) and is one of the common freeware processors that are currently available. Exemplarily, the DORIS processing chain (see figure 2.3) will be outlined in the following.

Before the interferogram can be formed according to eq. (2.4), the slave image is resampled to the master sampling grid. This requires an appropriate mapping function that is determined by a multi-stage coregistration (see section 2.3.2), the first stage of which is based on geocoding (see section 2.3.1). To reveal the signal of interest, the reference phase (2.6) is computed (see section 2.3.3) and subtracted from the interferometric phase. Finally, integer phase ambiguities can be resolved by phase unwrapping (see section 2.3.4). Optionally, some filtering operations can be performed to reduce noise in the interferometric phase (see section 2.3.5).

2.3.1. Geocoding

Geocoding means associating a pixel in the radar image with a position on the surface under consideration of the side-looking acquisition geometry. Besides its significance for data interpretation, geocoding is also required for a preliminary coarse coregistration (see section 2.3.2) and the reference phase computation (see section 2.3.3).

Starting from a pixel (ξ, η) in a SAR image, the location $\vec{x} = (x, y, z)^T$ of the corresponding surface target can be determined if its ellipsoidal height h is known. Given a position $\vec{x}_{\text{sat}}(\xi) = (x_{\text{sat}}, y_{\text{sat}}, z_{\text{sat}})^T$

on the satellite orbit, $\vec{x}$ can be computed by iteratively solving an equation system of three conditions (GEUDTNER, 1995, p. 84; DUT, 2009, ch. D.6):

- The target has a height h above a reference ellipsoid with the semiaxes a and b (ellipsoid condition):

$$\frac{x^2 + y^2}{(a+h)^2} + \frac{z^2}{(b+h)^2} - 1 = 0 \;. \tag{2.16}$$

- The distance to the target equals the measured range $R(\eta)$ (range condition):

$$|\vec{x}_{\text{sat}} - \vec{x}| - R(\eta) = 0 \;. \tag{2.17}$$

- The line of sight from the orbital position to the target is perpendicular to the satellite trajectory (zero-Doppler condition):

$$(\vec{x} - \vec{x}_{\text{sat}}) \cdot \dot{\vec{x}}_{\text{sat}} = 0 \;. \tag{2.18}$$

Note that the ellipsoid condition (2.16) is only an approximation – but a sufficiently accurate one. For elevations below 10 km, the absolute approximation error does not exceed 15 mm. The height h has to be iteratively interpolated from a DEM. If the terrain is rough, it is advisable to estimate timing errors δt and $\delta\tau$ from correlation of the DEM with a simulated amplitude image to mitigate their influence on the geocoding.

The geocoding procedure can also be applied inversely by deducing pixel coordinates (ξ, η) from a target position $\vec{x}$ and a given orbit trajectory.

2.3.2. Coregistration

Coregistration is required, because the sampling grids of master and slave image do generally not coincide. Significant differences in the acquisition start times t_1 and sampling window start times τ_1 can cause a constant shift of the image matrices. In general, even distortions of higher order can be observed, resulting from a lateral separation or an angular convergence of the two orbits. In order to resample the slave image to the master grid, a mapping polynomial of low degree d is required:

$$\begin{pmatrix} \xi_{\text{S}} \\ \eta_{\text{S}} \end{pmatrix} = \begin{pmatrix} \xi_{\text{M}} \\ \eta_{\text{M}} \end{pmatrix} + \sum_{i=0}^{d} \sum_{j=0}^{d-i} \begin{pmatrix} a_{ij} \xi_{\text{M}}^i \eta_{\text{M}}^j \\ b_{ij} \xi_{\text{M}}^i \eta_{\text{M}}^j \end{pmatrix} \;. \tag{2.19}$$

The estimation of polynomial coefficients proceeds in several stages. An initial coarse estimate of the coregistration shifts, i. e. the zero-order coefficients (a_{00}, b_{00}), can be obtained from orbit geometry: Assuming zero terrain height, a pixel $(\xi_{\text{M}}, \eta_{\text{M}})$ in the master image is geocoded, and a corresponding slave pixel $(\check{\xi}_{\text{S}}, \check{\eta}_{\text{S}})$ is computed by inverse geocoding in the slave geometry (see section 2.3.1). The resulting *coregistration offsets from orbits* or *orbital coregistration offsets*, respectively:

$$\begin{aligned} \Delta\check{\xi} &= \check{\xi}_{\text{S}} - \xi_{\text{M}} \\ \Delta\check{\eta} &= \check{\eta}_{\text{S}} - \eta_{\text{M}} \end{aligned} \tag{2.20}$$

are subsequently used to define the relative shift of homogeneously distributed pairs of patches from the two amplitude images. Cross-correlating these patches yields the more accurate *coregistration offsets from correlation*:

$$\begin{aligned} \Delta\xi &= \xi_{\text{S}} - \xi_{\text{M}} \\ \Delta\eta &= \eta_{\text{S}} - \eta_{\text{M}} \;, \end{aligned} \tag{2.21}$$

which indicate the difference in pixel coordinates of corresponding image features. This "coarse" correlation is further refined in another step using smaller correlation windows. Oversampling enables the determination of $\Delta\xi$ and $\Delta\eta$ on the subpixel level.

The offsets $(\Delta\breve{\xi}, \Delta\breve{\eta})$ and $(\Delta\xi, \Delta\eta)$ are in most cases identical within their level of accuracy. Only occasionally they exhibit significant differences resulting from either orbital effects or misregistration. As image resampling is exclusively based on correlation offsets $(\Delta\xi, \Delta\eta)$, it is insensitive to biases in orbital offsets $(\Delta\breve{\xi}, \Delta\breve{\eta})$. These biases mainly originate from relative timing errors but can also contain contributions from baseline errors or atmospheric propagation delay.

In many cases, the polynomial (2.19) can adequately describe the relative shift between individual pixels. A degree d of 2 or 3 may be sufficient for favourable conditions (NITTI ET AL., 2011), meeting a misregistration threshold of 1/8 pixel, below which the decrease of coherence is considered negligible (HANSSEN, 2001, p. 46). In cases with unfavourable conditions, i. e. rough topography, high elevation ranges, large perpendicular baselines or high resolution, it is advisable to additionally account for topographic effects. An appropriate algorithm, DEM-assisted coregistration, is described in (ARIKAN ET AL., 2008; DUT, 2009) and evaluated in (NITTI ET AL., 2011). In auroral regions it can be beneficial to additionally account for systematic azimuth offsets that result from ionospheric disturbances (WEGMÜLLER ET AL., 2006).

2.3.3. Reference Phase Computation

The reference phase ϕ_{ref} is defined in eq. (2.6) and can be obtained from the ranges R_M and R_S of master and slave acquisition to the geocoded target P (see figure 2.2). Given a pixel $(\xi_\text{M}, \eta_\text{M})$, the master position M is defined by the azimuth coordinate ξ_M, and P is determined by geocoding with $R_\text{M}(\eta_\text{M})$ as described in section 2.3.1. Finally, S is defined as the intersection of the slave orbit with its zero-Doppler plane through P, and R_S is the distance from S to P.

It is important to note that the image coordinates $(\xi_\text{S}, \eta_\text{S})$ of the corresponding slave pixel as defined by the coregistration polynomial (2.19) are not used for the computation of the reference phase. As a consequence, the slave timing information $(t_\text{S}, \tau_\text{S})$ is not required either. The reconstructed acquisition geometry of M, S and P rather implies a different set of image coordinates $(\breve{\xi}_\text{S}, \breve{\eta}_\text{S})$, where $\breve{\xi}_\text{S}$ and $\breve{\eta}_\text{S}$ are defined by S and R_S, respectively, using eq. (2.2). The discrepancy between $(\xi_\text{S}, \eta_\text{S})$ obtained from correlation and $(\breve{\xi}_\text{S}, \breve{\eta}_\text{S})$ obtained from orbits has already been discussed in the context of coregistration in section 2.3.2. In contrast to coregistration, the reference phase computation requires a consistent geometry and is thus relying on orbit information rather than correlation.

2.3.4. Phase Unwrapping

Phase unwrapping is indispensable for the estimation of relative displacements between the individual pixels of a SAR interferogram. A multitude of methods has been developed to infer absolute phases φ from the wrapped phase measurements (see GHIGLIA AND PRITT, 1998, for an elaborate introduction and an overview of basic concepts and difficulties; EINEDER AND HOLZNER, 1999; CHEN AND ZEBKER, 2001). Generally, approaches rely on the assumption that the phases of adjacent pixels do not differ by more than half a cycle. The actual challenge of phase unwrapping is that this assumption is sometimes violated, and a solution of maximum likelihood must be determined. Violations can be due to strong signal variations at scales smaller than the spatial sampling or phase noise in decorrelated patches (see section 2.2.4),. Incorrect unwrapping can bias deformation estimates very significantly and has thus to be avoided at any cost.

In some applications it is also feasible to infer the parameters of interest from the wrapped phase without explicitly resolving the ambiguities (FEIGL AND THURBER, 2009). For instance, if only a small number of geophysical fault parameters are to be estimated for the description of a coseismic tectonic process, knowledge of absolute phases is irrelevant for the final result. The parameters can be obtained by maximising an appropriate likelihood function. Nevertheless, such an approach does not actually avoid unwrapping, it should rather be considered as implicit unwrapping. Absolute phases can be calculated afterwards from the estimated parameters. However, these are optimised in a global sense and may contain some unlikely local artefacts.

In the context of time series approaches (see section 2.4), some three-dimensional methods for phase unwrapping have been developed (e. g., VAN LEIJEN ET AL., 2006; HOOPER AND ZEBKER, 2007; HOOPER, 2010). These additionally exploit the assumption that the phase of a coherent pixel does not change by more than half a cycle between two subsequent epochs. Such a comprehensive spatio-temporal unwrapping helps to connect disjoint patches that are separated by decorrelated regions in some interferograms.

2.3.5. Filtering

There are several approaches to reduce phase noise, which are applied at different stages of the processing chain (see figure 2.3).

Range Filtering It has already been pointed out in section 2.2.4 that the ground reflectivity spectrum of surface scatterers is shifted if the incidence angle θ_{inc} changes. This behaviour can be visualised by a different mapping of ground structures to the lines of sight of master and slave acquisition (see figure 2.4a). Hence, the object spectra of these structures are mapped to different portions of the data spectra (see figure 2.4b). But as only common parts of the object spectra can be exploited by interferometry, parts without counterpart in the spectrum of the other image rather create decorrelation noise than contribute to a coherent interferometric signal. Thus, it is possible to increase coherence by filtering out the non-common parts of the object spectra.

The design of an appropriate bandpass filter depends on the local fringe frequency (according to HANSSEN, 2001, p. 50, under consideration of eq. (2.1)):

$$f_\phi = \frac{1}{2\pi} \frac{\partial\phi}{\partial\tau} = -\frac{cB_\perp}{\lambda R \tan\left(\theta_{\mathrm{inc}} - \zeta\right)} \, , \tag{2.22}$$

which is identical with the relative frequency shift of the object spectra (see figure 2.4b). In case of relatively flat topography, f_ϕ can be computed from the orbit geometry, assuming a constant terrain slope, e. g., $\zeta = 0°$. If the terrain is rough and the interferometric coherence is sufficient, it is advisable to locally estimate the fringe frequency from the data and thus define individual filter parameters for small patches (HANSSEN, 2001, p. 49). For this adaptive procedure, a resampled slave image is a prerequisite, because a temporary interferogram needs to be computed for the estimation of the fringe frequency (DUT, 2009).

Azimuth Filtering In azimuth, there are two effects that can cause varying viewing directions: convergent orbit trajectories and different antenna squint angles or Doppler centroids, respectively. Although the mapping of the ground reflectivity spectrum is shifted in both cases (see figures 2.5a and b), the effects are not fully equivalent. During focussing, the data have already been filtered in azimuth to reduce noise

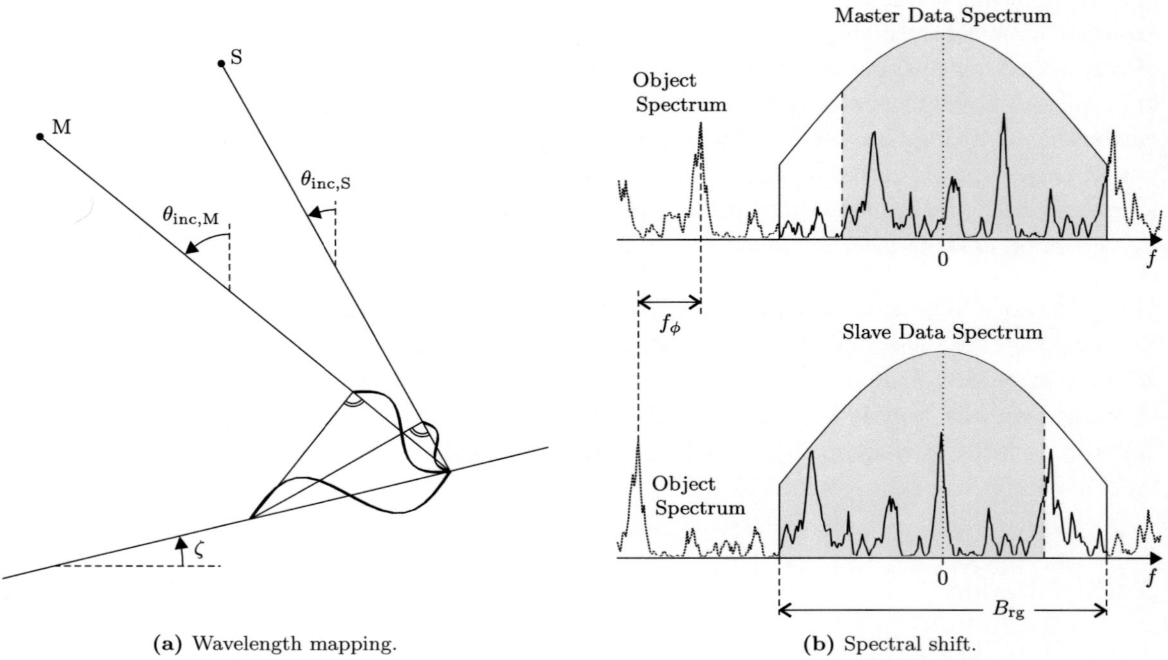

(a) Wavelength mapping.

(b) Spectral shift.

Figure 2.4.: Mapping of the ground reflectivity spectrum (object spectrum) in range. **(a)** A characteristic wavelength of a ground structure is mapped differently to the lines of sight of master and slave. (Figure reproduced from HANSSEN, 2001, figure 2.14 B) **(b)** The object spectrum is mapped to different portions of the data spectra. The overlapping part is determined by the chirp bandwidth B_{rg} and the fringe frequency f_ϕ. (Figure inspired by: GEUDTNER, 1995, figure 13; HANSSEN, 2001, figure 2.14 A)

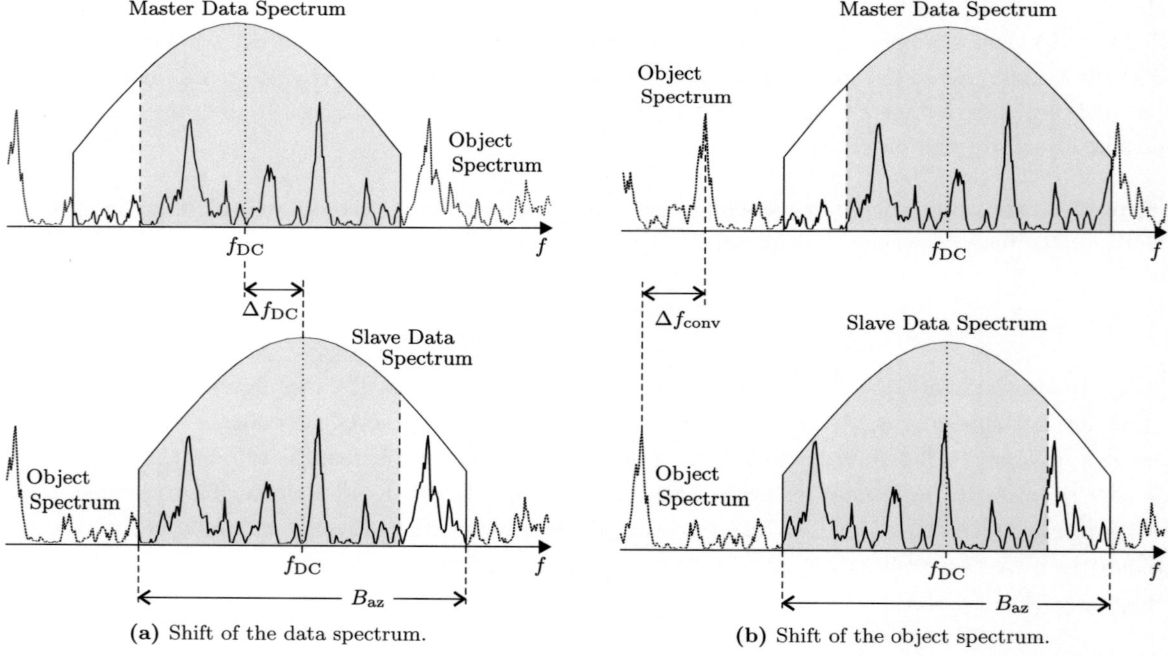

(a) Shift of the data spectrum.

(b) Shift of the object spectrum.

Figure 2.5.: Mapping of the ground reflectivity spectrum (object spectrum) in azimuth. (Figure inspired by GEUDTNER, 1995, figure 14) **(a)** Different portions of the data spectrum are isolated during focussing. Assuming parallel orbits, the overlapping part is determined by the processed Doppler bandwidth B_{az} and the respective Doppler centroid frequencies f_{DC}. **(b)** The object spectrum is mapped to different portions of the data spectra. Assuming coincident Doppler centroids, the overlapping part is determined by the processed bandwidth B_{az} and the frequency shift Δf_{conv}.

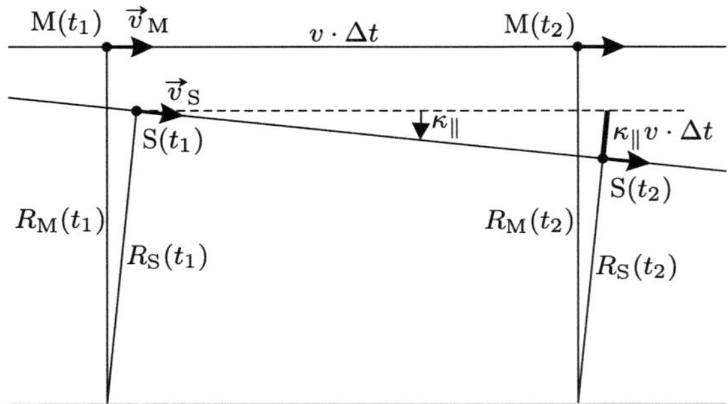

Figure 2.6.: Acquisition geometry with convergent orbits, projected onto a plane defined by the lines of sight of the master. Assuming an orbit convergence angle $\kappa_\parallel$ and constant target ranges for the master, the slave range decreases by a rate of $\kappa_\parallel v$.

and suppress ambiguities (BAMLER AND SCHÄTTLER, 1993, p. 93), and the passband has been centred at individual Doppler centroid frequencies f_DC. Hence, it is only the *difference* Δf_DC of Doppler centroids that causes decorrelation due to a narrowed overlapping portion of the data spectra (see figure 2.5a). An additional orbit convergence may either amplify or compensate this effect by introducing a relative shift Δf_conv of the object spectra (see figure 2.5b). This shift can be computed from the sensor velocity v and the 2D convergence angle $\kappa_\parallel$ (see figure 2.6 and eq. (3.9)) under consideration of eq. (2.6):

$$\Delta f_\mathrm{conv} = \frac{1}{2\pi} \frac{\partial}{\partial t} \left(-\frac{4\pi}{\lambda} (\Delta R_\mathrm{M} - \Delta R_\mathrm{S}) \right) = \frac{1}{2\pi} \frac{\partial}{\partial t} \left(-\frac{4\pi}{\lambda} \kappa_\parallel v \cdot \Delta t \right) = -\frac{2\kappa_\parallel v}{\lambda} \, . \tag{2.23}$$

Shifts in the Doppler centroid frequency are indeed an issue for some older missions like ERS or Radarsat. Filtering out non-common parts of the object spectra based on Δf_DC can significantly enhance coherence (SCHWÄBISCH AND GEUDTNER, 1995). During the Envisat mission, yaw-steering of the radar beam has improved significantly so that the decorrelation due to Δf_DC is almost negligible. This may also be the case for future missions. The effect of convergent orbits is also very small for standard acquisition scenarios and thus generally neglected (see also section 3.2).

The benefit of both azimuth and range filtering depends also on the specific application and the processing context. Filtering out non-overlapping parts of the object spectra enhances coherence only if a resolution cell is dominated by surface scatterers. For point scatterers, coherence is not sensitive to the viewing direction of the radar, and filtering would yield rather worsening than improvement.

Multilooking In the context of interferometric processing of SLC data, multilooking means enhancing the radiometric accuracy at the expense of geometric resolution by spatial averaging. With multilook factors (m_ξ, m_η) in azimuth and range, respectively, the multilooked interferogram reads:

$$\bar{z}_\mathrm{I}(\bar{\xi}, \bar{\eta}) = \frac{1}{m_\xi m_\eta} \sum_{(\xi,\eta) \in W} z_\mathrm{I}(\xi, \eta) \, , \tag{2.24}$$

where $W = \left\{ (\xi, \eta) : (\bar{\xi} - 1)m_\xi < \xi \le \bar{\xi} m_\xi \, \wedge \, (\bar{\eta} - 1)m_\eta < \eta \le \bar{\eta} m_\eta \right\}$. Multilooking is optimally applied to the complex interferogram after subtraction of the reference phase. By performing it before the reference phase subtraction, computation time can be saved at the cost of accuracy. This loss may be negligible for small multilook factors.

There are some limitations to the intensity of multilooking. If significant variations of small scale systematic signals (deformation or atmospheric) are averaged within one multilooked pixel, coherence may also degrade. Hence, there is always an optimal multilooking factor yielding maximum coherence for surface scatterers. Otherwise, multilooking rather deteriorates coherence of point scatterers, because these do not easily dominate an enlarged, multilooked resolution cell. Furthermore it is evident that the pixel size must not exceed the spatial wavelength of the signal of interest.

Phase Filtering   After subtraction of the reference phase, the distinctness of interferometric fringe patterns can be further enhanced to facilitate phase unwrapping. For this purpose, GOLDSTEIN AND WERNER (1998) proposed an adaptive filter that amplifies the dominant fringe frequencies in local patches, attenuating decorrelation noise at the same time. Evidently, this procedure involves small biases of the individual phases. However, these may have considerably less severe consequences than unwrapping errors and can be eliminated by restoring residuals after unwrapping.

2.4. Time Series Approaches

The InSAR deformation signal is generally superposed by atmospheric, orbital and topographic effects as well as noise (see eq. (2.7)). If the contribution of deformation is clearly dominant, viable deformation estimates can be obtained by simply neglecting other contributions at the expense of reduced sensitivity and accuracy. In order to better exploit the potential of the InSAR technique, an explicit distinction between signal components is indispensable. This is possible by either including complementary measurements like ground truth and meteorological data or considering correlation properties. The latter strategy is pursued by time series approaches, which are capable of inferring ground displacements as a function of time.

Whereas ground deformation can occur at all temporal and spatial scales, the characteristics of the associated interferometric signal ϕ_{defo} are filtered by the imaging system. Limitations in space are the size of a resolution cell and the width of the illuminated swath. The temporal dimension is confined by the revisit time of a sensor and the lifetime of a mission. But these restrictions may be overcome by combining acquisitions from different tracks or sensors.

Other signal components are usually separated from deformation by the correlation properties compiled in figure 2.7. The topographic phase error $\delta\phi_{\mathrm{topo}}$ can be identified by estimating the residual height error δh from its correlation with the perpendicular baseline $B_\perp$ (in equivalence with eq. (2.8)):

$$\delta\phi_{\mathrm{topo}} = -\frac{4\pi}{\lambda}\,\frac{B_\perp}{R\sin\theta_{\mathrm{inc}}}\,\delta h \ . \tag{2.25}$$

Both atmospheric and orbital effects may be considered uncorrelated for subsequent acquisitions with typical revisit intervals. Additionally considering their smoothness at small spatial scales, they can be identified by a combined spatial low-pass and temporal high-pass filtering. Atmospheric and orbital signals cannot be separated from each other, but this is not necessary either if deformation is the signal of interest. Finally, noise mitigation is most straightforward and can be achieved by spatial and/or temporal low-pass filtering.

A multitude of approaches have been developed that implement these concepts. There are two basic methodologies that can be considered fundamental for most current approaches, either dealing with *Persistent Scatterers* (PS) or limiting the processing to small baseline interferograms. They will be briefly outlined in the following subsections, focussing on their robustness with respect to orbit errors.

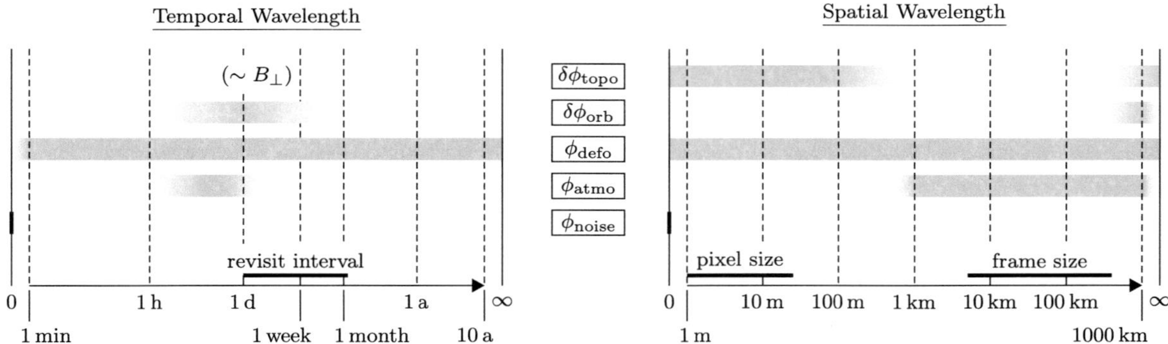

Figure 2.7.: Spatio-temporal correlation properties of the individual signal components listed in eq. (2.7). Topographic residuals ($\delta\phi_{\mathrm{topo}}$) can manifest in pixel-wise or local imperfections as well as global offsets or tilts; regional biases are considered rather unlikely. Whereas DEM errors δh are temporally constant, their effect on the phase is proportional to the perpendicular baseline $B_\perp$ (see eq. (2.8)). Temporal correlation of orbital effects ($\delta\phi_{\mathrm{orb}}$) has not been investigated in detail, but it is definitely disrupted by thrust maneuvers, which are performed at least every few weeks (see also section 3.2). Manifesting as an almost linear signal in interferograms, the spatial wavelength is far beyond the size of a radar scene. Ground deformation (ϕ_{defo}) may occur at all temporal and spatial scales. The atmospheric contribution (ϕ_{atmo}) can change completely within few hours, whereas variations are insignificant at spatial distances below 500 m. All remaining contributions, subsumed by ϕ_{noise}, do not show any correlation. Note that all signal components with wavelengths below twice the revisit interval or the pixel size, respectively, are aliased by the imaging system.

2.4.1. Persistent Scatterer InSAR

A very common family of InSAR time series approaches infers displacement estimates from a subset of permanently coherent pixels, the so-called *persistent scatterers* (PS). Their resolution cells are dominated by a point scatterer and are thus less affected by temporal decorrelation. The concept was initially introduced as *Permanent Scatterer* technique by FERRETTI ET AL. (2001), later refined (FERRETTI ET AL., 2000; COLESANTI ET AL., 2003) and seized by other research groups (ADAM ET AL., 2003; KAMPES, 2006; HOOPER ET AL., 2004, 2007; KETELAAR, 2009).

PS processing starts from a linearly independent set of interferometric combinations, all sharing one common master image. Pre-filtering in azimuth or range is not applied, because these operations are designed to reduce noise for distributed surface scatterers only. Multilooking and phase filtering are not useful either, since PS pixels are likely to be surrounded by pixels suffering from temporal decorrelation. For PS, these operations would rather promote signal decorrelation and thus enhance noise.

The processing chain strongly depends on the respective implementation and is often subject to continuous improvements. Hence, the initial approach of FERRETTI ET AL. (2001) differs in some relevant aspects from current methodologies. Nevertheless, three abstracted steps can be identified that are comprised by most implementations, eventually performed in combination or looped iteratively: PS selection, 3D unwrapping and *Atmospheric Phase Screen* (APS) estimation.

PS Selection Having processed the interferograms, PS pixels are identified by their phase stability, which is assessed by dedicated indicators. It is important to note that these indicators cannot be biased by orbit errors. They often operate with amplitude-based thresholding (FERRETTI ET AL., 2001; KAMPES, 2006; KETELAAR, 2009) that does not depend on the phase and is thus insensitive to atmospheric or orbital signals. So is the estimator proposed by HOOPER ET AL. (2004) and refined in (HOOPER ET AL., 2007), which analyses the phases of PS candidates after removing spatially correlated contributions. These contributions are estimated individually by applying a spatial band-pass filter to the phase of locally surrounding PS candidates. As the filter is applied in the frequency domain, it can easily cope with large phase gradients, eventually induced by strong orbit errors.

3D Unwrapping As ground displacement parameters can be estimated directly from the wrapped phase by dedicated estimators (FERRETTI ET AL., 2000; KAMPES AND HANSSEN, 2004), explicit phase unwrapping is not an essential requirement for PS processing. It was not part of the initial implementation by FERRETTI ET AL. (2001) either. However, later approaches generally include this step, because it enables the derivation of PS-wise displacement time series, which do not directly depend on a priori model assumptions. For unwrapping, different approaches can be pursued, either consecutively solving the ambiguities in the temporal and spatial dimension ("1D+2D" or "2D+1D", respectively; HOOPER ET AL., 2004; KAMPES, 2006) or providing a comprehensive 3D solution (VAN LEIJEN ET AL., 2006; HOOPER AND ZEBKER, 2007; HOOPER, 2010). Important prerequisites of successful unwrapping of PS phases are smallest possible spatial phase gradients, i. e., the phases of nearby PS should differ by less than π. If they do not, there are two conceivable remedies: Either the PS density is increased, or approximate estimates of systematic contributions are temporarily removed. Whereas the PS density cannot be influenced, PS approaches commonly mitigate the residual topographic and eventually the deformation signal by preliminary estimation and removal.

A very common approach is the consideration of phase differences corresponding to arcs that connect individual PS in a spatial network graph (FERRETTI ET AL., 2000; KAMPES, 2006; KETELAAR, 2009). Considering arcs has the advantage that the atmospheric and the orbital contribution can be neglected, because these are not expected to change significantly for short arc lengths. The relative height error $\Delta\delta h$, which also subsumes the sub-pixel offsets of the dominant scatterers, is estimated per arc by exploiting the correlation of $\delta\phi_{\text{topo}}$ with the perpendicular baseline $B_\perp$ (see eq. (2.25)). To consider the relative displacement, a parametric model like a linear rate of change or a seasonal oscillation is assumed.

Preliminary estimation of displacement parameters has the clear disadvantage that a functional model has to be postulated, chosen either from a priori knowledge or by statistically testing the performance of different models (VAN LEIJEN AND HANSSEN, 2007). Whereas preliminary displacement estimation is included in most PS processing chains, it is disapproved by HOOPER ET AL. (2004, 2007) who only estimate per PS an absolute height error δh (or look angle error, respectively) and a spatially uncorrelated contribution of the master image. Assuming the relative displacement of nearby PS to be small, their approach does not rely on the preliminary choice of a specific functional model. However, in contrast to other approaches, large spatial displacement gradients may be critical for successful unwrapping.

The most critical issue inherent to all unwrapping approaches is the validity of the assumption that phase differences of PS adjacent in time or space are smaller than π. Taking into account the unseizable noise contribution, it is rather desirable that systematic signal components differ by distinctly less than π. Preliminary mitigation of atmospheric and orbital signals is usually neither performed nor required, because their local gradients are mostly sufficiently small. However, even though this assumption applies to the great majority of applications, it might prove invalid in some cases with very large orbit errors, where only a preliminary estimation enables successful unwrapping.

APS Estimation The *Atmospheric Phase Screen* (APS) is an interferometric signal component that subsumes both the orbital and the atmospheric contribution, which are usually not separable from each other. Whereas some stochastic properties can be used to isolate parts of the atmospheric signal, it is impossible to reliably distinguish large scale atmospheric phase components from orbital effects without complementary measurements of the atmospheric state. It has thus become common practice to lump both contributions into the APS.

Considering the phases of individual PS, the APS is usually obtained by subsequently applying a temporal high-pass filter to the slave contribution and a spatial low-pass filter to the contributions of both master and slave. The master contribution is isolated by a temporal low pass filter beforehand (FERRETTI ET AL., 2000; HOOPER ET AL., 2007). Once the APS is estimated at PS locations, its interpolation to all pixel locations can be useful to reconsider the initial identification and selection of PS and iterate the processing.

The final processing result are time series of the interferometric phase of every PS. Having estimated and removed the contributions of residual topographic errors, atmosphere and orbits, the remaining phase can be considered as almost pure deformation signal, disturbed only by noise and eventual unwrapping errors (HOOPER ET AL., 2007). An appropriate functional displacement model can be evaluated and adjusted.

2.4.2. Small Baseline Approaches

Another common family of approaches is limited to processing only interferograms with small perpendicular and temporal baselines (BERARDINO ET AL., 2002; SCHMIDT AND BÜRGMANN, 2003) and focusses primarily on distributed scatterers that suffer from geometric and temporal decorrelation. As occurrences of these less persistent scatterers are often relatively dense compared to PS, they bear valuable information, which can also be enhanced by the filtering operations described in section 2.3.5. All interferograms that are successfully unwrappable in space may contribute, regardless of eventual linear dependencies with respect to other interferometric combinations. Due to filtering and unwrapping, which are both nonlinear operations, linearly dependent combinations are not fully redundant.

Having processed and unwrapped selected small baseline interferograms, a parametric model is adjusted to the individual phase values. Similar to PS processing, both a topographic and a displacement component may be considered. An essential innovation of the method is its applicability to multiple sets of interferograms formed from disjoint subsets of images (known as *small baseline subsets*, SBAS; BERARDINO ET AL., 2002) by computing a minimum norm solution.

By analogy to PS processing, spatial unwrapping can be supported by first estimating preliminary parameters, removing the model component and unwrapping the residuals. The parameters are then re-estimated after restoring the model component. The orbital signal is also lumped with the atmospheric phase, and both contributions are distinguished from deformation by subsequently applying a temporal high-pass and a spatial low-pass filter to the unwrapped phase. Even though the spatial sampling of coherent pixels is generally denser than for PS, the sensitivity of unwrapping with respect to large orbit errors may be an issue for ambiguity resolution across decorrelated patches.

2.4.3. Combined Approaches

PS and small baseline approaches are based on partly complementary sets of pixels. PS pixels are dominated by one point scatterer, and PS processing involves unfiltered interferograms with a single master. Most pixels considered by small baseline approaches are dominated by surface scattering, and interferograms are processed in various combinations of images, also involving filtering. Hence, in order to maximise the potential of InSAR time series analysis, both pixel groups have to be integrated into a joint processing. Such an approach has been proposed by HOOPER (2008) who embeds PS pixels into a processing chain based on small baseline interferograms. Conversely, FERRETTI ET AL. (2011) tackle the problem from the other side by integrating phase information from distributed scatterers into PS processing. Whereas the challenges related to unwrapping are mitigated due to an increased number of available observations, the basic problems inherent to PS and small baseline approaches remain the same.

2.5. Significance of Orbit Errors

This brief review of current InSAR methodology concludes with summarising why and in which respects imprecise knowledge of orbit trajectories may be critical for deformation monitoring. Three basic scenarios are distinguishable, in which the orbital error signal cannot be separated from the deformation signal and can thus be misinterpreted as deformation.

1. The most evident complication occurs when only one interferogram on its own is considered and orbit errors induce an almost linear phase ramp. Whereas a separation of signal components is not possible in this case, one out of three assumptions can be made:

   a) The deformation signal has a significantly larger magnitude than the orbital effect. Consequently, the orbital contribution is neglected, and an eventual orbital error signal is misinterpreted as deformation.

   b) The deformation signal has no global linear trend. An eventual trend is attributed to orbit errors and thus removed from the interferogram. Its removal also eliminates signal components related to large-scale deformation effects, but in presence of very large orbit errors this may be the lesser of two evils.

   c) The deformation signal is spatially confined, and the orbital signal outside the region of interest is representative for the whole scene. In this case, the orbital contribution is estimated from the assumed non-deforming regions and removed without distorting the deformation signal.

2. In time series approaches, orbital effects are mitigated by assuming that they are temporally uncorrelated. For large image stacks, they can be almost eliminated, whereas their isolation may be insufficient if too few images are available. The approach fails completely if the basic assumption did not apply. However, there is very little evidence for temporally correlated orbit errors, and if there is, the correlation might stem from hitherto undiscovered but predictable model errors (see section 3.4.3 and chapter 7).

3. Large orbit errors can complicate spatial unwrapping in presence of sparse spatial sampling, which is inherent to the PS approach and can also be conditioned by lacking coherence. Large orbital phase gradients may significantly contribute to a violation of the requirement that relative phase differences must be smaller than π. As a remedy, unwrapping can be supported by prior subtraction of an approximate estimate for the orbital contribution.

3. Orbits and Orbital Effects

Orbit errors are known to induce almost linear error signals in the interferometric phase. More precisely, these so-called phase ramps stem from biases in the reference phase caused by inaccurate knowledge of the interferometric baseline. The purpose of this chapter is to analyse the underlying mechanisms in detail and extend the investigations to other orbit-related effects like the convergence of trajectories, timing errors and clock errors.

Initially, the concept of the interferometric baseline is revisited in the context of a rigorously three-dimensional InSAR acquisition geometry. Subsequently, the long-term behaviour of the baseline is analysed exemplarily for the Envisat mission in order to draw some conclusions regarding orbit convergence. The quantitative aspect of orbit errors is addressed in section 3.3 where methodology, limitations and accuracy of precise orbit determination (POD) are reviewed. Finally, a detailed error analysis is conducted in order to identify individual error patterns in the interferometric phase and coregistration offsets and quantify them under consideration of their driving mechanisms.

Simulations of phase error patterns require the assumption of an acquisition geometry and thus always refer to a specific mode of a specific sensor. For more than a decade, acquisition parameters of the ERS satellites, which are very similar to those of the Envisat's *Image Swath 2* (IS2), have been considered an implicit standard for simulations and analysis of InSAR-specific effects (e. g., HANSSEN, 2001). Hence, all numbers and error patterns presented in this chapter are based on Envisat parameters and conferrable to the ERS.

A couple of new missions have been launched recently (see table 3.1), providing a multitude of different acquisition modes. To account for this recent and future variety, implications to selected other missions are additionally mentioned, and some numerical details are given in appendix A. Besides ERS-1/2 and Envisat, this selection comprises the planned sequel mission Sentinel-1 as well as Radarsat-1/2, which are the only operational C-band missions after the failure of Envisat in April 2012. The *Advanced Land Observing Satellite* (ALOS) and TerraSAR-X complete the selection as sample missions for L-band and X-band.

3.1. The Interferometric Baseline

The idealised InSAR acquisition geometry in figures 2.2 and 3.1a, respectively, assumes that trajectories of repeated satellite passes are rigorously parallel, which is generally not correct. In reality, the orbits of two overflights are to some degree convergent (see figure 3.1b), implying that the plane spanned by the points M, S and P is perpendicular neither to the master nor to the slave orbit. The assumption of a rigorously three-dimensional geometry has consequences for the decomposition of the interferometric baseline, which is defined here as the difference vector of the sensor positions M and S:

$$\vec{B}(t, R) = \vec{x}_\mathrm{S}(t, R) - \vec{x}_\mathrm{M}(t) \, . \tag{3.1}$$

Whereas the variation of the baseline vector in azimuth (timing $t := t_\mathrm{M}$) is obvious, its dependence on range R is often disregarded as it is relatively weak. There is indeed no dependence for parallel orbits,

Table 3.1.: Overview of past, current and planned spaceborne SAR missions that acquire images on a regular basis and are dedicated (at least partly) for civilian use. The respective operation time refers to the period in which SAR images are or have been acquired on a regular basis, beginning with the end of the commissioning phase. (For Cosmo-Skymed 4, the author was unable to retrieve this date.) ΔT is the orbit repeat cycle of a single satellite regardless its constellation. The highlighted missions are receiving special attention within the scope of this thesis. (ASI, 2007; CSA, 2012; COVELLO ET AL., 2009; DLR, 2013; ESA, 2012, 2013; GANTERT ET AL., 2011; GEUDTNER ET AL., 2011; GÓMEZ ET AL., 2012; HANSSEN, 2001; IMPAGNATIELLO ET AL., 1998; KANKAKU ET AL., 2009; KRIEGER ET AL., 2007; LAUR, 2011; MDA, 2013; MEZZASOMA ET AL., 2008; MOREIRA ET AL., 2011; JAXA, 2008, 2012; SAUNIER ET AL., 2007; SCHARROO, 2002; WERNINGHAUS AND BUCKREUSS, 2010; ZINK ET AL., 2011)

Mission (Sensor)	Operator	Band	λ [cm]	Launch	Operation	H [km]	ΔT [d]
ERS-1	ESA (EU)	C	5.66	1991	1991–2000	790	3/35/168
JERS-1	NASDA (J)	L	23.5	1992	1992–1998	568	44
ERS-2	ESA (EU)	C	5.66	1995	1995–2011	790	3/35
Radarsat-1	CSA (CA)	C	5.66	1995	1995–2013	800	24
ENVISAT (ASAR)	ESA (EU)	C	5.62	2002	2003–2012	790	30/35
ALOS-1 (PALSAR-1)	JAXA (J)	L	23.6	2006	2006–2011	700	46
Cosmo-Skymed-1	ASI (I)	X	3.1	2007	2008–	620	16
TerraSAR-X	DLR (D)	X	3.11	2007	2008–	514	11
Cosmo-Skymed-2	ASI (I)	X	3.1	2007	2008–	620	16
Radarsat-2	MDA (CA)	C	5.55	2007	2008–	800	24
Cosmo-Skymed-3	ASI (I)	X	3.1	2008	2009–	620	16
TanDEM-X	DLR (D)	X	3.11	2010	2010–	514	11
Cosmo-Skymed-4	ASI (I)	X	3.1	2010		620	16
planned missions:							
Sentinel-1a	ESA (EU)	C	5.55	2014		700	12
PAZ	Hisdesat (E)	X	3.11	2014		514	11
ALOS-2 (PALSAR-2)	JAXA (J)	L	23.8	2014		628	14
Sentinel-1b	ESA (EU)	C	5.55	2015		700	12
TerraSAR-X2	DLR (D)	X	3.11	2016		514	11
Radarsat Constellation	CSA (CA)	C	5.55	2018		593	12
Tandem-L	DLR (D)	L	23.8	2019		760	8

but the baseline varies over range if the orbits are convergent. Due to the zero-Doppler condition (2.18), two targets P_1 and P_2 that are acquired from the same point M on the master orbit do not necessarily share a common acquisition point S on the slave orbit.

Compared to its expression by coordinate differences in a geocentric frame, a more convenient representation of the baseline is given in a moving frame that is aligned to the master orbit. A common realisation of such a frame is a Frenet frame, which is characterised by the following unit vectors in along-track, radial (**vertical**) and across-track (**horizontal**) direction:

$$\vec{e}_a = \frac{\vec{v}_M}{|\vec{v}_M|}, \qquad \vec{e}_v = \frac{\vec{x}_M}{|\vec{x}_M|} - \left\langle \frac{\vec{x}_M}{|\vec{x}_M|}, \vec{e}_a \right\rangle \vec{e}_a, \qquad \vec{e}_h = \vec{e}_a \times \vec{e}_v, \tag{3.2}$$

where $\langle \cdot, \cdot \rangle$ is the dot product (inner product) and $\vec{v}_M$ the velocity vector of the master satellite. The baseline decomposition in the Frenet frame of the master orbit $(\vec{e}_h, \vec{e}_a, \vec{e}_v)$ reads:

$$\begin{aligned} B_h &= \left\langle \vec{B}, \vec{e}_h \right\rangle \\ B_a &= \left\langle \vec{B}, \vec{e}_a \right\rangle \\ B_v &= \left\langle \vec{B}, \vec{e}_v \right\rangle . \end{aligned} \tag{3.3}$$

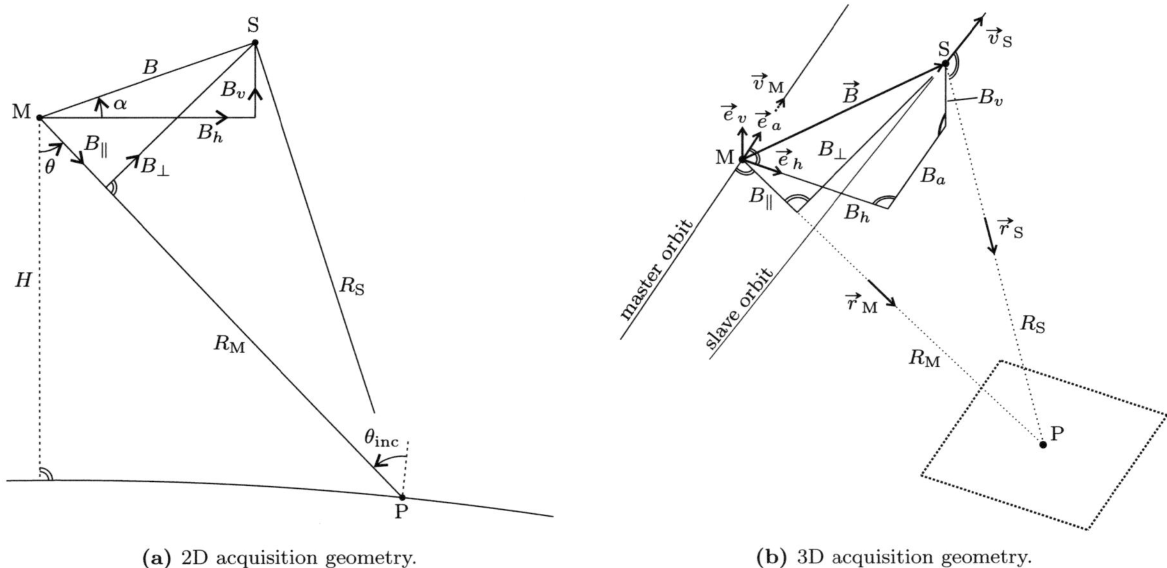

(a) 2D acquisition geometry.

(b) 3D acquisition geometry.

Figure 3.1.: InSAR acquisition geometry: A target P on the surface is acquired from point M on the master orbit and point S on the slave orbit. **(a)** Idealised geometry assuming perfectly parallel orbits, viewed in a plane perpendicular to the flight direction. **(b)** Actual geometry with convergent orbits.

For interferometric purposes, the baseline component $B_\perp$ perpendicular to the line of sight is decisive for the height sensitivity of the geometric phase in eq. (2.8) and establishes an alternative representation together with the associated parallel component $B_\parallel$:

$$B_\parallel = \left\langle \vec{B}, \vec{r}_{\mathrm{M}} \right\rangle$$
$$B_\perp = \sqrt{\left|\vec{B}\right|^2 - B_\parallel^2} \,, \tag{3.4}$$

where $\vec{r}_{\mathrm{M}}$ is a unit vector aligned to the line of sight of the master acquisition. As the convergence of orbits is indeed very small, it is common to approximate these components by their projections onto the cross-track plane (see figure 3.1a):

$$B_\parallel \approx B_h \sin\theta - B_v \cos\theta$$
$$B_\perp \approx B_h \cos\theta + B_v \sin\theta \,, \tag{3.5}$$

assuming $B_a \approx 0$. This simplification is very helpful for theoretical considerations, but can entail more or less significant biases when used for the computation of the height ambiguity (2.8). A third conventional representation of the baseline is given by its length B and its orientation angle α (HANSSEN, 2001, p. 117):

$$B = \left|\vec{B}\right| \approx \sqrt{B_h^2 + B_v^2}$$
$$\alpha(t, R) = \arctan\frac{B_v}{B_h} \,. \tag{3.6}$$

It follows:

$$\begin{aligned} B_h &\approx B\cos\alpha & B_\parallel &\approx B\sin(\theta - \alpha) \\ B_v &\approx B\sin\alpha & B_\perp &\approx B\cos(\theta - \alpha) \,. \end{aligned} \tag{3.7}$$

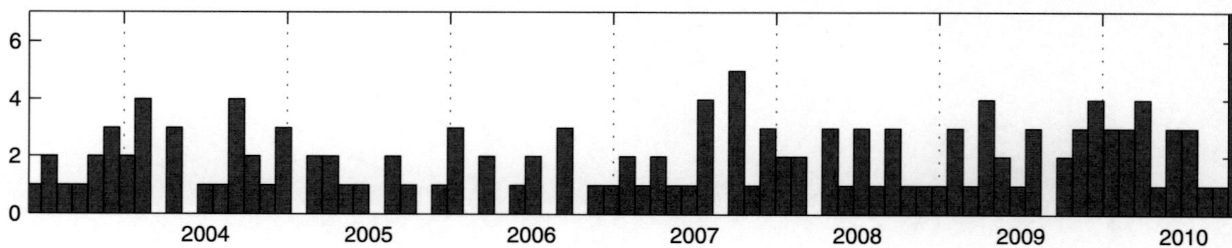

Figure 3.2.: Number of orbit calibration maneuvers per 35-day Envisat revisit cycle between 1 June 2003 and 24 Oct 2010 (data source: ESOC Envisat homepage, `http://nng.esoc.esa.de/envisat/`, accessed on 14 Feb 2012).

The mutual convergence of master and slave orbit is characterised by the convergence angle:

$$\kappa = \arccos\left\langle \frac{\vec{v}_M}{|\vec{v}_M|}, \frac{\vec{v}_S}{|\vec{v}_S|} \right\rangle , \tag{3.8}$$

which is further decomposable into two components $(\kappa_\parallel, \kappa_\perp)$ reflecting the rates of change of parallel and perpendicular baseline. $\kappa_\parallel$ is defined as the projection of κ onto the $(\vec{r}_M, \vec{e}_a)$-plane (see figure 2.6) and quantifies the geometric decorrelation of interferograms due to orbit convergence (see section 2.3.5). With $\frac{\vec{v}_M}{|\vec{v}_M|} = \vec{e}_a$, $\frac{\vec{v}_S}{|\vec{v}_S|} := \vec{v}_{S,0}$ and the normal vector $\vec{r}_M \times \vec{e}_a$ of the projection plane it reads:

$$\kappa_\parallel = \frac{\dot{B}_\parallel}{v} = \arccos\left\langle \vec{e}_a, \vec{v}_{S,0} - \langle \vec{v}_{S,0}, \vec{r}_M \times \vec{e}_a \rangle (\vec{r}_M \times \vec{e}_a) \right\rangle . \tag{3.9}$$

$\kappa_\perp$ is the projection onto the $(\vec{r}_M \times \vec{e}_a, \vec{e}_a)$-plane and is useful to assess the accuracy of the baseline approximation (3.5). With the plane's normal vector $\vec{r}_M$, it is defined by:

$$\kappa_\perp = \frac{\dot{B}_\perp}{v} = \arccos\left\langle \vec{e}_a, \vec{v}_{S,0} - \langle \vec{v}_{S,0}, \vec{r}_M \rangle \vec{r}_M \right\rangle . \tag{3.10}$$

3.2. The Envisat Orbit

Until October 2010, Envisat followed a near-circular, sun-synchronous orbit with an altitude between 785 and 814 km. One revolution took 100.6 minutes, and the ground track had a repeat cycle of precisely 35 days, which is also known as revisit interval. As the orbital plane is subject to secular drifts driven by perturbing forces, orbit maintenance maneuvers were carried out in irregular intervals. Concerted hydrazine thruster burns ensured that the satellite maintained within a 1 km wide tube defined by a conventional reference trajectory (DUESMANN AND BARAT, 2007). From late 2010 on, the tight orbit control was abandoned to save combustibles at the expense of limited interferometric capability.

Figure 3.2 gives an impression of the frequency of orbit maneuvers. Maintenance intervals were ranging from several burns a day to maneuver-free periods of more than two months. Consequently, it is the rule rather than the exception that at least one maneuver took place between two acquisitions of the same scene, which are temporally separated by integer multiples of 35 days. This circumstance motivates an important conclusion for the temporal correlation of orbit errors: Even if there would be correlation to some degree for subsequent acquisitions, which is difficult to assess, the correlation would definitely be disrupted by interfering maneuvers. Hence, it appears justified to assume that orbit errors of subsequent acquisitions are uncorrelated.

In order to provide a reasonably general impression of the potential magnitude of orbit convergence, a comprehensive baseline analysis has been carried out (see figure 3.3). For the complete, continuous orbit

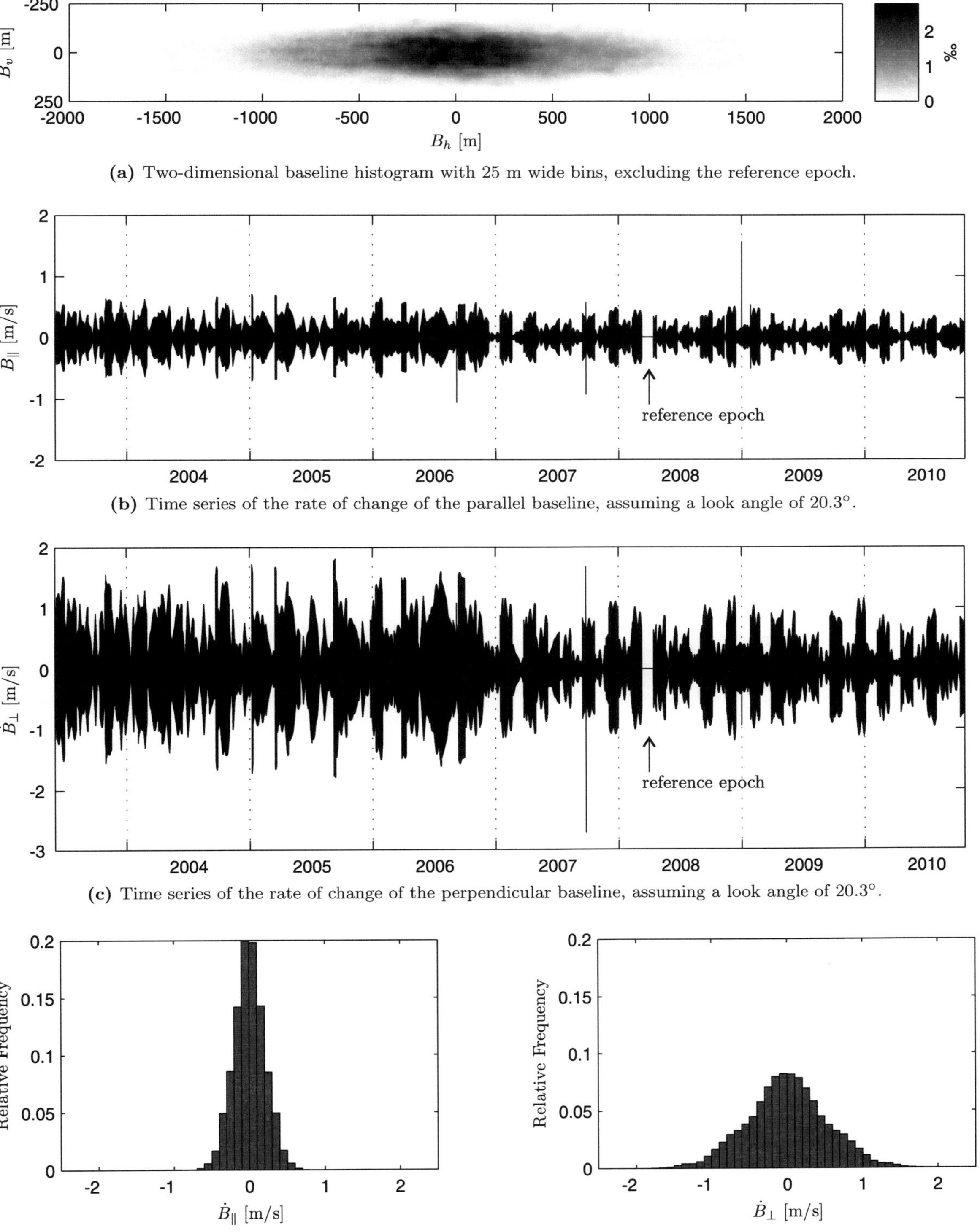

(a) Two-dimensional baseline histogram with 25 m wide bins, excluding the reference epoch.

(b) Time series of the rate of change of the parallel baseline, assuming a look angle of $20.3°$.

(c) Time series of the rate of change of the perpendicular baseline, assuming a look angle of $20.3°$.

(d) Histogram of the values in subfigure b, excluding the reference epoch.

(e) Histogram of the values in subfigure c, excluding the reference epoch.

Figure 3.3.: Baseline analysis of the complete Envisat orbit trajectory between 1 June 2003 and 24 Oct 2010, not distinguishing between periods of actual SAR acquisitions and inactivity of the ASAR instrument. The baseline has been computed with respect to a reference trajectory, for which the maneuver-free reference epoch between 10 Mar and 14 Apr 2008 has been chosen.

trajectory between 1 June 2003 and 24 Oct 2010, the horizontal and vertical baseline components B_h and B_v have been computed with respect to a reference trajectory, assuming $B_a = 0$. The histogram in figure 3.3a gives an impression of the orbital tube. It appears broader than the specified width of 1 km, which is to be maintained only at the equator though, whereas the histogram covers orbit data from all latitudes between $\pm 81.5°$. Furthermore, the arbitrarily chosen reference trajectory does not necessarily coincide with the conventional reference orbit defined by the European Space Agency (ESA), which may also let the tube appear broader.

From the rates of change $\dot{B}_h$ and $\dot{B}_v$, the rates $\dot{B}_\parallel$ and $\dot{B}_\perp$ have been inferred (see figures 3.3b and c) using the approximation (3.5). The assumed look angle of 20.3°corresponds to mid-range of the conventional *Image Swath 2* (IS2). Whereas the rates themselves are arbitrary as they describe the orbit convergence with respect to the reference trajectory, the difference of the rates from two overflights is an indicator for the orbit convergence κ between the respective acquisitions as defined in eqs. (3.8) through (3.10). From $\kappa_\perp$, the approximation error of computing $B_\perp$ with eq. (3.5) can be inferred:

$$\delta B_\perp = \kappa_\perp B_\perp \ . \tag{3.11}$$

Assuming a pessimistic value of $\dot{B}_\perp = 2$ m/s ($\kappa_\perp = 0.27$ mrad $= 0.015°$; see figure 3.3e) yields only an insignificant error of $\delta B_\perp = 2.7 \cdot 10^{-4} B_\perp$. $\kappa_\parallel$, on the other hand, is a measure for the geometric decorrelation due to convergent orbits (see section 2.3.5). The associated frequency shift follows from eqs. (2.23) and (3.9):

$$\Delta f_\mathrm{conv} = -\frac{2 \dot{B}_\parallel}{\lambda} \ . \tag{3.12}$$

Assuming a pessimistic value of $|\dot{B}_\parallel| = 1$ m/s ($|\kappa_\parallel| = 0.13$ mrad $= 0.008°$; see figure 3.3d), the absolute frequency shift would amount to 36 Hz and imply a decorrelation of 2.7 % according to eq. (2.13) with the processed Doppler bandwidth $B_\mathrm{az} = 1316$ Hz. Considering that this value has been computed with most pessimistic assumptions, it may be concluded that geometric decorrelation due to convergent orbits is hardly significant for Envisat.

This conclusion may be extended to the ERS satellites, the orbits of which have likewise been maintained within a 1 km wide tube (DUESMANN AND BARAT, 2007). The effect will be even less pronounced for later missions with a tighter orbit control like TerraSAR-X (500 m wide tube, YOON ET AL., 2009) or Sentinel-1 (100 m wide tube, GEUDTNER ET AL., 2011). However, it may become significant for the Radarsat-1 satellite, the horizontal orbit variations of which span even 5 km (VACHON ET AL., 1995). Furthermore, it is notable that $\dot{B}_\parallel$ and thus $\kappa_\parallel$ increases with the look angle due to the oblate shape of the orbital tube (see figure 3.3a). Hence, the contribution of orbit convergence to geometric decorrelation would have to be reassessed for shallow looking acquisition modes.

3.3. Precise Orbit Determination

Precise Orbit Determination (POD) is a complex procedure based on measurements by space geodetic techniques and on complementary force models. The resulting orbit is usually given in terms of state vectors that describe the trajectory of a satellite at regularly sampled points in time. State vectors comprise at least three position coordinates, eventually a three-dimensional velocity vector and they may be extended by a number of additional model parameters (SCHARROO, 2002, p. 51). The latter play an important role in POD but are meaningless for interferometric processing though.

Table 3.2.: Available *high precision* techniques for Precise Orbit Determination (POD). For Radarsat-1, only radar tracking data with inferior precision are available.

Platform	Launch	Altitude	High Precision Techniques
ERS-1	1991	790 km	SLR, altimetry
ERS-2	1995	790 km	SLR, PRARE, altimetry
Radarsat-1	1995	800 km	
Envisat	2002	790 km	SLR, DORIS, altimetry
ALOS	2006	700 km	GPS, SLR
TerraSAR-X	2007	514 km	GPS, SLR
Radarsat-2	2007	800 km	GPS
Sentinel-1a	2014	963 km	GPS

3.3.1. Space Geodetic Techniques

The following space geodetic techniques are and have been used for POD of SAR satellites (see table 3.2):

SLR *Satellite Laser Ranging* deduces the distance from a ground station to a satellite by measuring the travel time of a laser pulse. The pulse is transmitted from the ground station, reflected by an on-board retro-reflector and received at the same ground station. This relatively expensive technique requires some manual effort in targeting and tracking individual satellites. It also has a limited capacity, because a ground station can only track one satellite at a time. Additionally, SLR suffers from weather restrictions and the irregular distribution of the tracking stations, yielding considerable quality variations of orbit products (SCHARROO, 2002, pp. 50 and 53). Another limitation can be imposed by optical sensors on-board multi-sensor platforms (e. g., ALOS) that can be damaged by the laser when operating during tracking (NAKAMURA ET AL., 2007).

Altimetry Radar altimeters measure the height of a satellite above the sea surface by evaluating the round-trip travel time of a downwardly transmitted radar pulse. Although this technique has not been designed for orbit determination in the first place, altimetric measurements can be used to support the radial component of the orbit if the required sea-level heights are predicted from oceanographic models (SCHARROO AND VISSER, 1998). The risk that oceanic signals leak into the orbits is limited to spatial scales larger than 10 000 km and may be significantly outweighed by the gain in stability of the orbit solution (SCHARROO, 2002, pp. 64-66).

PRARE The German *Precise Range And Range-Rate Equipment* (BEDRICH, 1998) operates with dual-frequency microwave signals that are transmitted by the satellite, repeated by a ground-based transponder and received again by the satellite. PRARE makes code-based range measurements and additionally deduces the range rate from the Doppler shift of the carrier phase. In the late 1990s, the PRARE ground segment qualified by a more homogeneous distribution than the SLR tracking network (MASSMANN ET AL., 1997). To date, the PRARE system is no longer operational.

DORIS *Doppler Orbitography and Radiopositioning Integrated by Satellite* (JAYLES ET AL., 2006) is a French development that also operates with dual-frequency microwave signals. These are transmitted by ground-based beacons and received by the satellite. Similar to PRARE, the range-rate is obtained from the Doppler shift of the carrier frequency. The DORIS tracking network stands out due to an excellent global coverage.

GNSS *Global Navigation Satellite Systems* enable a continuous positioning by dual-frequency pseudo-range and carrier phase measurements to several medium earth orbit GNSS satellites at a time. So far, only the US-American *Global Positioning System* (GPS) has been used for POD of SAR satellites. In terms of temporal and spatial coverage, GPS outperforms any of the aforementioned techniques. On the other hand, orbit determination from GPS measurements is an indirect procedure, because it depends on the GPS orbits, which have to be determined themselves beforehand.

3.3.2. Methodology

A closed representation of a satellite orbit by six Keplerian parameters is not feasible at a high level of accuracy. A purely analytical description is considerably complicated by a multitude of perturbing forces (MONTENBRUCK AND GILL, 2000, ch. 3): oblateness of the earth and higher order inhomogeneities of the gravity field, gravitational forces of other celestial bodies (moon, sun, planets), earth and ocean tides, solar radiation pressure, atmospheric drag, thrust forces of orbit maintenance maneuvers, terrestrial radiation pressure (albedo) and relativistic effects.

Methods of numerical POD can be subdivided into *dynamic* and *kinematic* approaches. Dynamic orbits are obtained by adjusting all measurements to a combined model of perturbed Keplerian motion. The estimation problem is usually partitioned into multiple overlapping arcs, each spanning several days of tracking data. Dynamic approaches can cope with sparse tracking data but require an initial, approximate solution. Kinematic orbits on the other hand are obtained from adjusting measured positions to an arbitrary smooth curve. This requires continuous tracking data, which are only available from GNSS. Nevertheless, the practical use of kinematic methods is significantly restricted by their sensitivity to outliers, unfavourable viewing geometry and data gaps (MONTENBRUCK ET AL., 2004).

For dynamic POD, a number of models and assumptions is required to reproduce geometry and mechanics of the satellite revolution (SCHARROO, 2002, p. 61): measurement corrections, ground station coordinates, gravity models, tidal displacement, third body attraction, atmospheric drag, radiation pressure, orbit maneuvers, the reference frame and a satellite model including mass cross-sections. Especially atmospheric drag forces, which are very significant for low earth orbiting satellites, are not easily predictable. A good remedy are so-called *reduced-dynamic* approaches, which extend the parameter set by empirical accelerations. Their application is quite common for POD of SAR satellites.

There are different stages of orbit products, reflecting increasing levels of precision subject to the respective delay of their release. For ERS and Envisat, the ESA distinguishes between predicted, restituted, preliminary and precise orbits (CLOSA, 1998), whereas different terminologies may be employed for other missions, e. g., "precise rapid" and "precise science" orbits (YOON ET AL., 2009) or "rapid science" and "precise science" orbits (BOCK ET AL., 2011). Waiting some days with orbit processing enables constraining the dynamic model by including measurements of the later course of the trajectory. An even longer processing delay does not apply a different estimation model but rather benefits from the availability of more or more precise intermediate products of input data, more accurate ancillary information and the opportunity to invest more time in solving problems (SCHARROO, 2002, pp. 59 et seq.). In general, only precise or scientific orbits, respectively, are of interest for InSAR applications. Merely in case of short term disaster management, the usage of less accurate orbit products is worth considering.

3.3.3. Accuracy Assessment and Validation

Assessing the influence of inaccurate orbits on interferometric products requires some insight into the challenges of orbit accuracy validation and the informative value of generic quality measures. SCHARROO

(2002, p. 64) lists the following potential indicators: tracking data residuals, altimeter crossovers, collinear track differences and arc overlap differences. Additionally, the impact of using different force models or processing strategies can be evaluated. However, none of these approaches can provide a general accuracy estimate for various reasons.

A rigorous validation by tracking data residuals is only possible if the data in question have not been used for the orbit determination itself. Otherwise, only a lower bound for the actual orbit error can be provided (OTTEN AND DOW, 2005). Altimeter crossover differences are qualified to infer gravity model errors, because crossing orbital arcs are sensitive to spatially separate parts of the gravity field. Contrariwise, collinear tracks experience identical gravity model errors and are thus only affected by non-gravitational error sources. Consecutive arcs or solutions from different analysis centres rely at least partly on the same data, and their comparison cannot be considered a rigorously independent validation. Approaches involving altimetric measurements are only sensitive to radial orbit errors. And finally, systematic errors in modelling the atmospheric propagation delay affect all measurements in the same way, even if independent techniques are involved.

Regardless the availability of independent data sets, it is noteworthy that global quality measures are not necessarily stationary in time and space. There may be local quality variations due to an inhomogeneous distribution of SLR tracking stations. An increased solar activity or unfavourable atmospheric conditions can entail less accurate solutions, and in some cases validation campaigns comprise only selected epochs, which are not necessarily representative for the whole mission.

Another difficulty consists in the variety of ways to define quality measures. Mostly, *root mean squares* (RMS) of orbit differences or tracking residuals, respectively, are given, but also standard deviations, maximum deviations or ranges of residuals may be offered. For these values, it is rarely distinguished between along-track, across-track and radial components; often only the radial or a total 3D error is given. These variants are not compatible with the requirements of SAR interferometry though, where only relative errors (i. e., baseline errors) have a notable effect and geographically correlated errors stemming from gravity models can safely be ignored. Additionally, interferograms are almost insensitive to baseline errors in the along-track component, and even in the cross-track plane the sensitivity is distinctly anisotropic (see section 4.3.1). Hence, considering the across-track orbit error, which is always larger than the radial error, as an upper bound to baseline errors is often the only practicable approach to evaluate the impact of orbit errors on InSAR applications.

It may be concluded that inferring the orbit accuracy for a particular SAR acquisition from a global quality indicator works fine in the majority of cases but does not account for sporadic outliers. In the following, the conclusions from orbit accuracy assessments for the different missions are summarised, explicitly avoiding a direct comparison because of the just illustrated complexity.

ERS-1/2 The *European Remote Sensing Satellites* were not designed for highly accurate orbit determination (SCHARROO AND VISSER, 1998). Nevertheless, owing to the coexistence of an altimeter instrument on the same platform, a considerable effort has been undertaken in POD from SLR and altimeter measurements (SCHARROO AND VISSER, 1998; SCHARROO, 2002; DOORNBOS AND SCHARROO, 2005). Whereas additional inclusion of PRARE data for ERS-2 was considered an improvement in the 1990s (VISSER ET AL., 1997), this technique has later been disregarded due to the progressive degradation of the tracking network (DOORNBOS AND SCHARROO, 2005). In view of the primary importance of altimetric applications, accuracy assessments focus on the radial component. The RMS of 5 cm initially obtained by SCHARROO AND VISSER (1998) could later be enhanced to 4 cm for ERS-2 orbits (DOORNBOS AND SCHARROO, 2005) due to the availability of new gravity models from the *Gravity Recovery and Climate Experiment* (GRACE). Though suggested in many publications, SCHARROO AND VISSER (1998) did *not* provide a rigorously independent evaluation of the across-track component. From arc overlaps and comparisons of different solutions its accuracy may be estimated to be in the order of 8 cm or worse (VISSER ET AL., 1997; SCHARROO AND VISSER, 1998; OTTEN ET AL., 2011). Due to the inhomogeneous quality of SLR-derived orbits, occasional outliers can be observed more frequently than for more recent missions relying on DORIS or GPS.

Envisat   Compared to ERS, orbit products for Envisat significantly benefit from the additional observations of the DORIS instrument. Altimetric measurements can be used for a rigorously independent validation of the radial component. OTTEN AND DOW (2005) assessed the accuracy of the best solutions to be 3 cm radially and 10 cm in 3D, continuously improving towards the solar minimum in late 2008.

Radarsat-1   The design of the first Radarsat satellite did not consider InSAR applications at all. In contrast to ERS-1/2, no altimeter is part of the payload either, which would have implied more exigent requirements for orbit determination. As a consequence, Radarsat-1 operates in a 5 km wide orbital tube, is not yaw-steered, and the orbit is determined by less precise ground-based radar tracking (VACHON ET AL., 1995; GEUDTNER ET AL., 1998; S. Côté, CSA, pers. comm., 2012). There has been only one validation campaign that involved ground-based transponders and the detection of their responses in SAR images (CSA, 2010). Due to the poor accuracy of these measurements, only an upper bound for the actual orbit error of a few tens of metres can safely be inferred. PEPE ET AL. (2011) give a heuristic estimate of the orbit accuracy based on InSAR applications, which is "on the order of some meters". Besides, the relatively loose orbit control, which implies less frequent calibration maneuvers, may also explain occasional temporal correlations of orbit errors, which have been observed by (HOOPER ET AL., 2007, p. 11).

Radarsat-2   Compared to its predecessor, orbit maintenance and control has significantly improved for Radarsat-2. The platform is yaw-steered, and GPS receivers are available for orbit determination (MORENA ET AL., 2004). Whereas radar tracking can be provided in case of failure of GPS, there is no independent technique available for validation purposes. Hence, the orbit accuracy is inferred from the geolocation accuracy of known point targets in high resolution imaging products and estimated to be better than 10 m (M. Chabot, MDA, pers. comm., 2012). Baseline errors of some sample interferograms have been observed in the order of several decimetres up to more than a metre.

ALOS   The *Advanced Land Observing Satellite* was the first SAR satellite to be tracked continuously with a GPS receiver, using SLR only for validation purposes. However, this advantage was relativised by electromagnetic interference of the GPS signals with the *Phased Array type L-band SAR* (PALSAR) and eventually other payload instruments (NAKAMURA ET AL., 2007). It resulted in frequent loss of GPS signals and an underperforming orbit determination accuracy. NAKAMURA ET AL. (2007) characterise SLR residuals of an independent validation campaign by a systematic offset of -4.4 cm and a standard deviation of 6.6 cm. They further show RMS deviations of daily arc overlaps that are between 2 and 15 cm, maximum deviations exceeding 30 cm in some cases. This suggests an actual accuracy of a few decimetres, which is still significantly below the official requirement specification of 1 m (ESA, 2007b).

TerraSAR-X   TerraSAR-X orbits are exclusively based on GPS observations. YOON ET AL. (2009) concluded from an independent validation by SLR that the "Precise Science Orbits" have a 3D RMS accuracy of 2 cm. As this number suffers from some of the aforementioned general limitations of validation, they also give a heuristic estimate based on various validation approaches, according to which the unbiased 3D RMS accuracy should be "definitely better than 10 cm".

3.4. Error Mechanisms

In InSAR processing, orbit data are required for geocoding master and slave image, i. e., identifying the location of the individual resolution cells on the ground. The thus obtained positions are used for two processing steps: coarse coregistration and reference phase computation.

Coarse coregistration from orbits is no strict requirement for processing. It produces an approximate initial offset value for coregistration from correlation, which is only essential in case of very large timing errors. Contrariwise, coregistration from orbits and thus the orbits themselves can be validated with the correlation approach. Validation should be possible at an accuracy of $1/20$ pixel under most optimistic conditions (HANSSEN, 2001, p. 46), which corresponds to several decimetres for standard C-band missions. However, this accuracy is usually not achieved in practice.

The reference phase on the other hand is distinctly more sensitive to orbit errors. Biases are measurable on the millimetre level but difficult to isolate from atmospheric and deformation signals. In contrast to the effect of orbit errors on coregistration, these biases directly affect the final deformation estimates, which is why their study is of primary interest. The following orbit error analysis thus focusses on patterns in the interferometric phase φ but will also cover coregistration offsets from orbits $(\Delta\check{\xi}, \Delta\check{\eta})$ as defined in eq. (2.20).

Different types of orbit-related errors can be distinguished: baseline errors $\delta\vec{B}$ and errors $(\delta t, \delta\tau)$ in the annotated timing t in azimuth and τ in range (see figure 3.4). Their detailed analysis will be extended to clock errors, which affect the radar carrier frequency and the sampling rates. For a number of error parameters, corresponding error patterns will be presented for both interferometric phase and orbital coregistration offsets. To enable a quantitative assessment, it will also be specified in particular, how large an error in baseline, timing or frequency needs to be that the resulting error pattern spans a range of 2π or induces maximum offsets of 1 pixel, respectively. The following simulations and computations will assume a spherical earth body. Further details on model assumptions and sensor parameters can be found in appendix A.

3.4.1. Baseline Errors

The limitation to consider only baseline errors instead of absolute orbit errors stems from the circumstance that absolute errors are negligible if they apply in the same way to both master and slave. This is evident from the definition of the reference phase in eq. (2.6), for which the following approximation is valid:

$$\phi_{\mathrm{ref}} = -\frac{4\pi}{\lambda}\left(R_{\mathrm{M,ref}} - R_{\mathrm{S,ref}}\right) \approx -\frac{4\pi}{\lambda}B_{\parallel} . \tag{3.13}$$

The approximation by the parallel baseline is sufficiently accurate for general theoretical considerations. It was introduced by ZEBKER AND GOLDSTEIN (1986), evaluated by HANSSEN (2001, p. 67 et seq.) and is also known as *plane wave approximation* (BAMLER AND HARTL, 1998; RICHARDS, 2009), *parallel-ray approximation* (ZEBKER ET AL., 1997) or *far field approximation* (HANSSEN, 2001). Analogous conclusions can be drawn for biases of the the orbital coregistration offsets from eq. (2.20) by expressing them as a function of baseline errors $\delta\vec{B}$ only. With the pulse repetition frequency f_{PRF} and the range sampling rate f_{RSR} from eq. (2.2) follows:

$$\delta\Delta\check{\xi} \approx \frac{f_{\mathrm{PRF}}}{v}\left(\delta B_a + R\,\delta\kappa_{\parallel}\right) = \frac{f_{\mathrm{PRF}}}{v}\left(\delta B_a + \frac{R}{v}\,\delta\dot{B}_{\parallel}\right) \approx \frac{f_{\mathrm{PRF}}R}{v^2}\,\delta\dot{B}_{\parallel}$$
$$\delta\Delta\check{\eta} = \frac{2f_{\mathrm{RSR}}}{c}\left(\delta R_{\mathrm{S}} - \delta R_{\mathrm{M}}\right) \approx -\frac{2f_{\mathrm{RSR}}}{c}\,\delta B_{\parallel} . \tag{3.14}$$

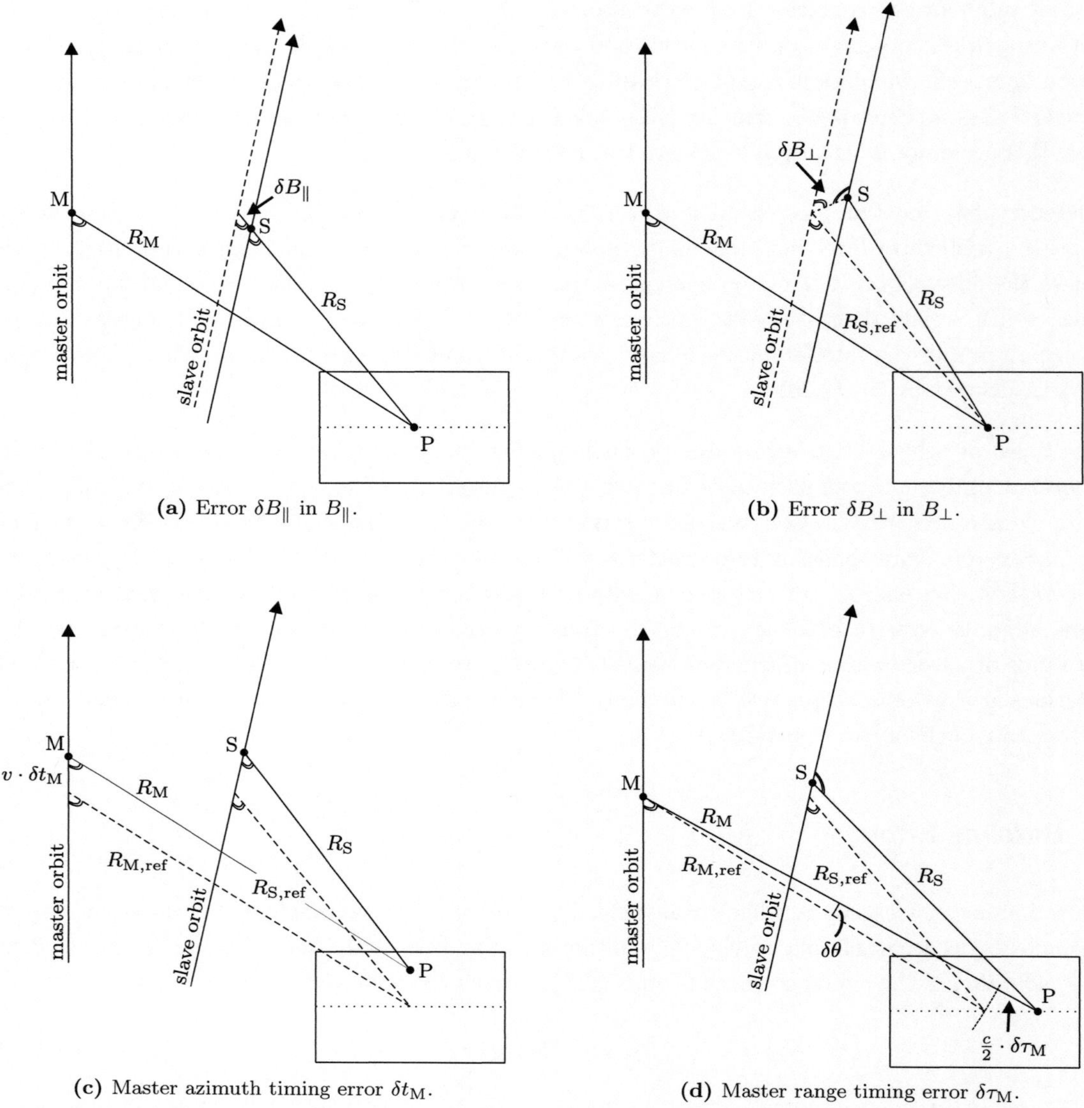

(a) Error $\delta B_{\parallel}$ in $B_{\parallel}$.

(b) Error $\delta B_{\perp}$ in $B_{\perp}$.

(c) Master azimuth timing error δt_{M}.

(d) Master range timing error $\delta\tau_{\mathrm{M}}$.

Figure 3.4.: Geometric interpretation of baseline errors and timing errors in the context of reference phase computation. **(a-b)** These visualisations assume a deterministic master orbit. **(c-d)** Slave timing errors have no effect on the reference phase.

In order to evaluate the sensitivity of an interferogram to baseline errors, errors $\delta\varphi$ in the residual phase φ as defined in eq. (2.7) are now considered as a function of an error baseline $\delta\vec{B}$ that is superposed to the interferometric baseline. Taylor series expansion in azimuth time t and range R at the scene centre (t_0, R_0) yields under consideration of $\partial B_\parallel / \partial\theta = B_\perp$ and $\partial B_\perp / \partial\theta = -B_\parallel$ (see eq. (3.5)):

$$
\begin{aligned}
\delta\varphi(t, R) = {} & -\delta\phi_{\text{ref}}(t, R) \\
= {} & \frac{4\pi}{\lambda} \Big\{ \delta B_\parallel(t_0, R_0) + \delta\dot{B}_\parallel(t_0, R_0)\, dt + \delta B_\perp(t_0, R_0)\, \frac{\partial\theta}{\partial R}(R_0)\, dR \\
& + \frac{1}{2}\delta\ddot{B}_\parallel(t_0, R_0)\, dt^2 + \delta\dot{B}_\perp(t_0, R_0)\, \frac{\partial\theta}{\partial R}(R_0)\, dt\, dR \\
& + \frac{1}{2}\left[\delta B_\perp(t_0, R_0)\, \frac{\partial^2\theta}{\partial R^2}(R_0) - \delta B_\parallel(t_0, R_0)\left(\frac{\partial\theta}{\partial R}(R_0)\right)^2 \right] dR^2 + \ldots \Big\}.
\end{aligned}
\tag{3.15}
$$

The dependence of θ on R is determined by the individual topography of an acquired scene. Partial derivatives for a spherical earth surface, which is assumed here, are given in eq. (A.1). For coregistration offsets in azimuth and range, the respective Taylor expansions read:

$$
\begin{aligned}
\delta\Delta\breve{\xi}(t, R) = {} & \frac{f_{\text{PRF}}}{v^2} \Big\{ \delta\dot{B}_\parallel(t_0, R_0)\, R_0 + \delta\ddot{B}_\parallel(t_0, R_0)\, R_0\, dt \\
& + \left(\delta\dot{B}_\parallel(t_0, R_0) + \delta\dot{B}_\perp(t_0, R_0)\, R_0\, \frac{\partial\theta}{\partial R}(R_0) \right) dR + \ldots \Big\} \\
\delta\Delta\breve{\eta}(t, R) = {} & -\frac{2f_{\text{RSR}}}{c} \Big\{ \delta B_\parallel(t_0, R_0) + \delta\dot{B}_\parallel(t_0, R_0)\, dt + \delta B_\perp(t_0, R_0)\frac{\partial\theta}{\partial R}(R_0)\, dR \\
& + \frac{1}{2}\delta\ddot{B}_\parallel(t_0, R_0)\, dt^2 + \delta\dot{B}_\perp(t_0, R_0)\, \frac{\partial\theta}{\partial R}(R_0)\, dt\, dR \\
& + \frac{1}{2}\left[-\delta B_\parallel(t_0, R_0)\left(\frac{\partial\theta}{\partial R}(R_0)\right)^2 + \delta B_\perp(t_0, R_0)\, \frac{\partial^2\theta}{\partial R^2}(R_0) \right] dR^2 + \ldots \Big\},
\end{aligned}
\tag{3.16}
$$

assuming a constant satellite velocity v.

Based on the Taylor series in eqs. (3.15) and (3.16), the influence of baseline errors $\delta\vec{B} = (\delta B_\parallel, \delta B_\perp)$ and their time derivatives on either the phase or coregistration can be evaluated. A convenient quantitative characterisation is given by the amount of error in an individual baseline parameter that is required to produce a maximum phase difference of 2π within one scene or a maximum coregistration offset of 1 pixel, respectively. These error numbers are labelled by the respective baseline parameters ($\delta B_\parallel$, $\delta B_\perp$, $\delta\dot{B}_\parallel$ and $\delta\dot{B}_\perp$) and subscripted by "2π" for the phase, "ξ" for the coregistration offset in azimuth and "η" for the coregistration offset in range. They are determined by isolated evaluation of the respective most significant Taylor coefficients, yielding:

$$
\begin{array}{lll}
\delta B_{\parallel, 2\pi} = -\dfrac{4\lambda}{(\Delta\theta)^2} & \delta B_{\parallel, \xi} = 0 & \delta B_{\parallel, \eta} = -\dfrac{c}{2f_{\text{RSR}}} \\[2ex]
\delta B_{\perp, 2\pi} = \dfrac{\lambda}{2\,\Delta\theta} & \delta B_{\perp, \xi} = 0 & \delta B_{\perp, \eta} = -\dfrac{c}{f_{\text{RSR}}\,\Delta\theta} \\[2ex]
\delta\dot{B}_{\parallel, 2\pi} = \dfrac{\lambda}{2\,\Delta t} & \delta\dot{B}_{\parallel, \xi} = \dfrac{v^2}{f_{\text{PRF}}R} & \delta\dot{B}_{\parallel, \eta} = -\dfrac{c}{f_{\text{RSR}}\,\Delta t} \\[2ex]
\delta\dot{B}_{\perp, 2\pi} = \dfrac{\lambda}{\Delta t\,\Delta\theta} & \delta\dot{B}_{\perp, \xi} = \dfrac{2v^2}{f_{\text{PRF}}R\,\Delta\theta} & \delta\dot{B}_{\perp, \eta} = -\dfrac{2c}{f_{\text{RSR}}\,\Delta t\,\Delta\theta}
\end{array}
\tag{3.17}
$$

with $\Delta\theta = \frac{\partial\theta}{\partial R}\Delta R$. Numerical values for Envisat IS2 can be found in figure 3.5, where also the corresponding error signals have been simulated. Note that these signals depend neither qualitatively nor quantitatively on the absolute value of the perpendicular baseline $B_\perp$.

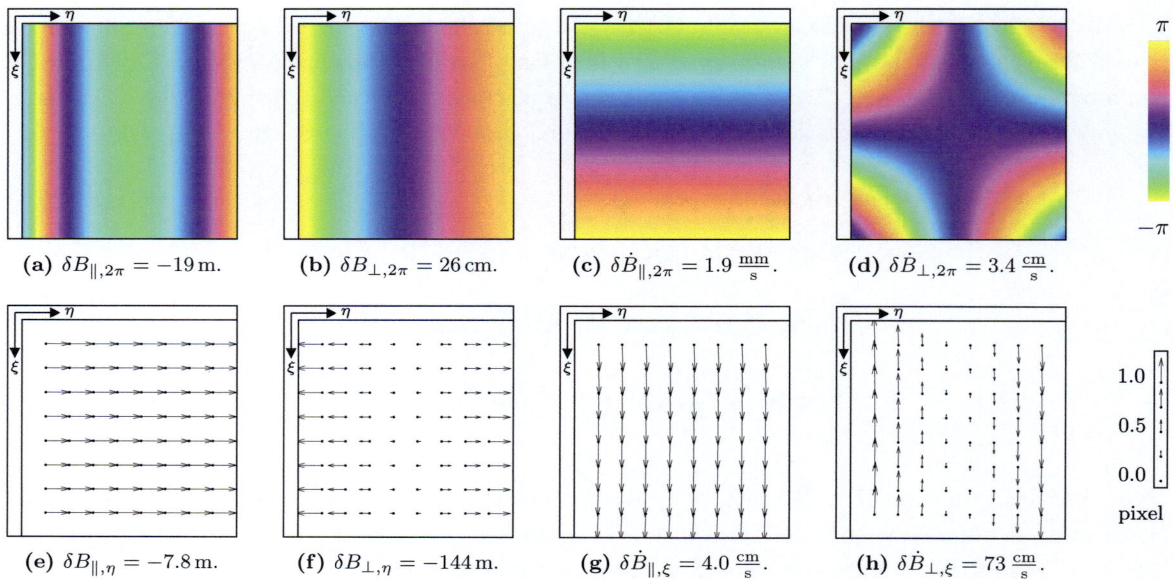

Figure 3.5.: Patterns in the phase and orbital coregistration offsets (2.20), respectively, that are induced by errors in the baseline and its time derivatives into an Envisat IS2 interferogram with a spherical reference surface. All patterns show a maximum phase difference of 2π or a maximum coregistration offset of 1 pixel, respectively. The baseline errors that are required to induce these patterns have been computed from eq. (3.17). The baseline decomposition into $B_\parallel := B_\parallel(R_0)$ and $B_\perp := B_\perp(R_0)$, which is for parallel orbits a function of range, is defined by mid-range R_0 according to eq. (A.5). **(a-d)** Error patterns in the interferometric phase. **(e-h)** Coregistration offset patterns, computed with $f_{\mathrm{PRF}} = 1652$ Hz and $f_{\mathrm{RSR}} = 19.21$ MHz. The effect of errors in $\dot{B}_\parallel$ and $\dot{B}_\perp$ on range offsets ($\delta\dot{B}_{\parallel,\eta} = -1.04$ m/s, $\delta\dot{B}_{\perp,\eta} = -19$ m/s) is obscured by their largely stronger effect on azimuth offsets.

From the phase patterns and the corresponding baseline errors in figures 3.5a-d can be observed that errors in $B_\perp$ and $\dot{B}_\parallel$ have by far the most significant effect on the phase. An error in $B_\perp$ of 26 cm induces an almost linear ramp in range (figure 3.5b), and an error in $\dot{B}_\parallel$ of 1.9 mm/s induces an almost linear ramp in azimuth (figure 3.5c). A superposition of these two elementary signals can result in a ramp with any orientation (see figure 3.6). Errors in $B_\parallel$ and $\dot{B}_\perp$ may be considered insignificant, because they require a notably larger amount of error to have a visible effect on the interferogram. Even if one of these error components was very large, the associated effect would probably be obscured by contributions of $\delta B_\perp$ and $\delta\dot{B}_\parallel$.

An error in $B_\parallel$, for instance, is expected to translate into a homogeneous phase shift, which changes only the colour but not the shape of the interferometric fringe pattern. Hence, there should be no error pattern at all. The fringe pattern in figure 3.5a, however, results from the dependence of the baseline decomposition $(B_\parallel, B_\perp) = (B_\parallel(R), B_\perp(R))$ on range. As the pattern has been simulated with a fixed decomposition defined by the line of sight to the centre of the scene, an error $\delta B_\parallel(R_0)$ leaks into the perpendicular component for $R \neq R_0$, inducing a small phase variation over range.

The significance of errors in $B_\perp$ is supported by the assessment of orbit accuracy in section 3.3.3. With an absolute 3D RMS of 10 cm, the relative accuracy perpendicular to the line of sight, which is the relevant component for InSAR applications, may be assumed to be at the level of some centimetres. Hence, a measurable error signal of a fraction of a fringe is expectable in the standard case and may exceed one fringe in some exceptional cases. For errors in $\dot{B}_\parallel$, the assessment of significance is more difficult, because directional errors of the satellite's heading are usually not covered by publications on orbit validation. There is merely one indication by SCHARROO AND VISSER (1998) who mention a slope error of 0.2 μrad $\hat{=}$ 1.4 mm/s for the Delft ERS orbits. Even though this is rather an absolute than a relative measure

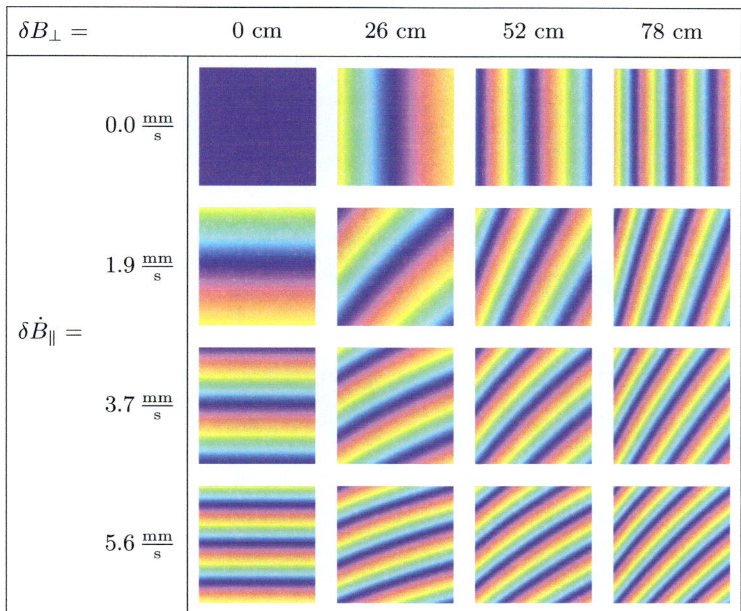

Figure 3.6.: Superposition of error signals due to $\delta\dot{B}_{\parallel}$ and $\delta B_{\perp}$ for Envisat IS2, simulated for integer numbers of fringes in azimuth and range, respectively.

and applies to a different satellite, it allows at least to validate the general significance of errors in $\dot{B}_{\parallel}$ considering that one fringe is induced by $\delta\dot{B}_{\parallel,2\pi} = 1.9$ mm/s.

Computations of the sensitivity indicators $\delta B_{\parallel,2\pi}$, $\delta B_{\perp,2\pi}$, $\delta\dot{B}_{\parallel,2\pi}$ and $\delta\dot{B}_{\perp,2\pi}$ for other sensors than Envisat's *Advanced SAR* (ASAR) can be found in table A.3. Based on the orbit quality survey in section 3.3.3, the effect of baseline errors on interferograms may be assessed more or less significant for current missions, most for Radarsat-1 and least for TerraSAR-X. For the latter, baseline errors can be considered almost negligible.

The annotations to figures 3.5e-h suggest that coregistration offsets are considerably less sensitive to baseline errors than the interferometric phase. Keeping in mind that orbital offsets can be validated by image correlation with a most optimistic accuracy of 1/20 pixel (HANSSEN, 2001, p. 46), inference of Envisat baseline errors from coregistration can never perform more accurate than 4 dm in $B_{\parallel}$, 4 m in $B_{\perp}$, 2 mm/s in $\dot{B}_{\parallel}$ or 2 cm/s in $\dot{B}_{\perp}$, respectively. This level is easily outperformed by baseline error estimation from the interferometric phase, which will be discussed in detail in chapter 4.

Whereas the phase patterns in figures 3.5a-d are simulated with a spherical reference surface, the orbital error signal is indeed sensitive to variations of the topographic height. Nonetheless, this sensitivity only applies to errors in $B_{\perp}$ and is only significant for large height differences. Some examples are shown in figure 3.7, where a phase ramp of four fringes in range (figure 3.7a) is distorted by the topography of different generic DEM (figures 3.7b-d). The resulting nonlinear features have important implications for the parameterisation of baseline errors, which will be discussed in section 4.4.3.

3.4.2. Timing Errors

Even though errors in the timing annotations of SAR image data cannot be considered orbit errors in the strict sense, they are closely related and thus complementarily covered here. It can be distinguished between azimuth timing errors δt and range timing errors $\delta\tau$.

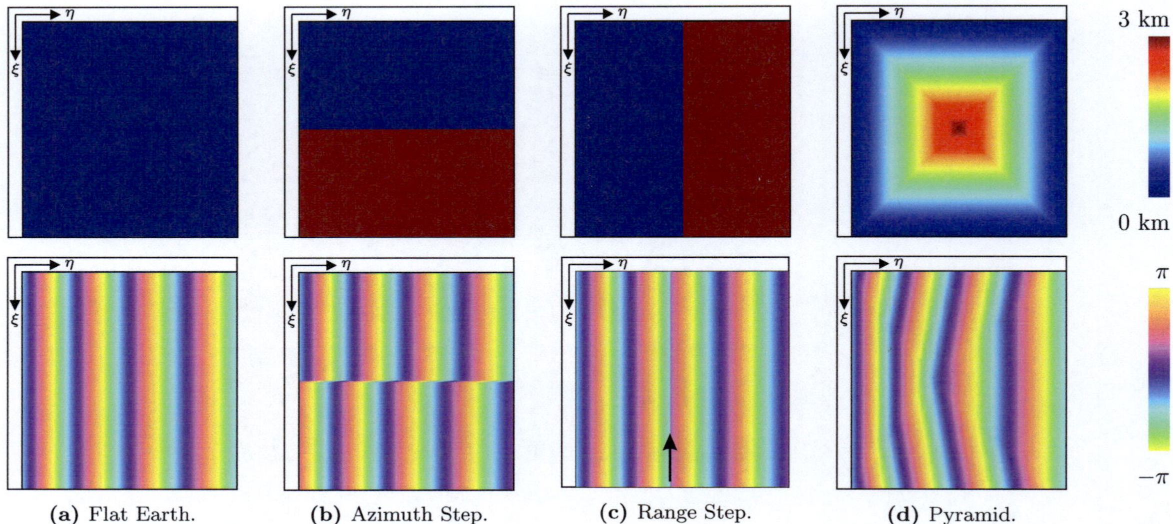

(a) Flat Earth.　　**(b)** Azimuth Step.　　**(c)** Range Step.　　**(d)** Pyramid.

Figure 3.7.: Error patterns in the interferometric phase that are induced by a baseline error of $4\,\delta B_{\perp,2\pi} = 1.05$ m (corresponding to four fringes in range), assuming different generic DEM. The upper row shows the respective DEM and the lower row the corresponding error signals. Note that this effect is independent of the absolute baseline $B_{\perp}$.

Azimuth timing indicates the absolute acquisition time t of all pixels sharing a common zero-Doppler plane (see figure 2.1b); errors therein are basically synchronisation biases of the on-board clock with respect to terrestrial clocks. As orbit determination relies on terrestrial clocks, azimuth timing errors cause that the point of acquisition is mapped to an incorrect position on the orbit. Equivalent effects can result from along-track orbit errors, which are not separable from azimuth timing errors. Range timing is the measured two-way travel time τ of the radar pulse from the sensor to the target and back. Apparent errors therein result either from the time measurement itself or from inaccurate or lacking modelling of the atmospheric propagation delay.

The timing errors discussed here are assumed to remain constant during the acquisition. Hence, a delayed azimuth timing is equivalent to an anticipated acquisition start time t_1, and an extended signal travel time is equivalent to a smaller sampling window start time τ_1. The following investigations consider thus only two timing error parameters: $\delta t \equiv -\delta t_1$ and $\delta \tau \equiv -\delta \tau_1$. Eventual timing error drifts rather result from a biased pulse repetition frequency or range sampling rate, respectively, and are thus covered by the discussion of clock errors in section 3.4.3.

It can also be distinguished between absolute and relative timing errors. An absolute error δt_{M} in the azimuth timing of the master can be interpreted geometrically by a shift of the assumed acquisition point M on the master orbit with respect to its actual position (see figure 3.4c). An absolute error $\delta \tau_{\mathrm{M}}$ in the range timing of the master results in a shifted position of the target P on the surface (see figure 3.4d). As the slave acquisition point S is determined from P by inverse geocoding (see section 2.3.1), slave timing is not required to reconstruct the acquisition geometry and thus disregarded for the reference phase computation by the DORIS InSAR processor. However, slave timing errors $(\delta t_{\mathrm{S}}, \delta \tau_{\mathrm{S}})$ do have an effect on orbital coregistration offsets. As master timing errors with an opposite sign have very similar effects, coregistration cannot be considered sensitive to slave timing errors themselves but rather to relative timing errors between master and slave. Hence, the following investigations will cover the effects of master timing errors $(\delta t_{\mathrm{M}}, \delta \tau_{\mathrm{M}})$ on the reference phase and the effect of relative timing errors:

$$\delta \Delta t = \delta t_{\mathrm{S}} - \delta t_{\mathrm{M}}$$
$$\delta \Delta \tau = \delta \tau_{\mathrm{S}} - \delta \tau_{\mathrm{M}} \tag{3.18}$$

on coregistration offsets.

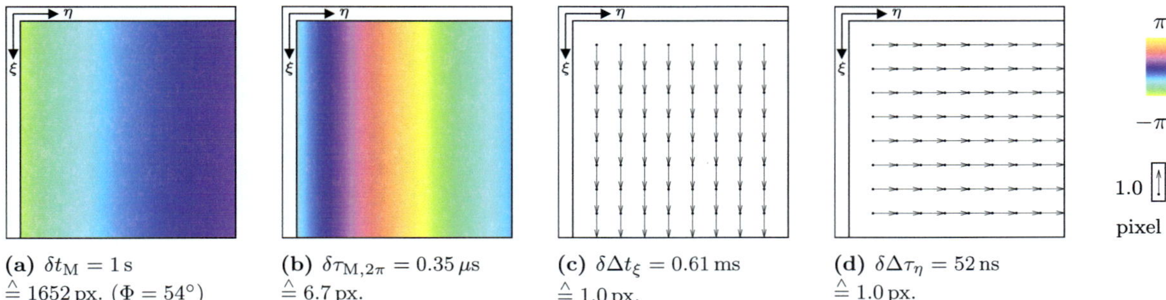

(a) $\delta t_{\mathrm{M}} = 1\,\mathrm{s}$
$\overset{\wedge}{=} 1652\,\mathrm{px.}$ $(\Phi = 54°)$

(b) $\delta\tau_{\mathrm{M},2\pi} = 0.35\,\mu\mathrm{s}$
$\overset{\wedge}{=} 6.7\,\mathrm{px.}$

(c) $\delta\Delta t_{\xi} = 0.61\,\mathrm{ms}$
$\overset{\wedge}{=} 1.0\,\mathrm{px.}$

(d) $\delta\Delta\tau_{\eta} = 52\,\mathrm{ns}$
$\overset{\wedge}{=} 1.0\,\mathrm{px.}$

Figure 3.8.: Patterns in the phase and orbital coregistration offsets (2.20), respectively, that are induced by timing errors into an Envisat IS2 interferogram with $B_{\parallel} = 0\,\mathrm{m}$ and $B_{\perp} = 500\mathrm{m}$. **(a)** Phase pattern simulated with an ellipsoidal reference surface at 54° latitude of the WGS 84 ellipsoid for a very large azimuth timing error. The orientation of the phase ramp depends on the local curvature of the ellipsoid and thus on latitude. The effect is practically insignificant, even weaker at other latitudes and usually dominated by terrain-related effects of the timing error. **(b)** Phase pattern simulated with a flat spherical reference surface, showing exactly one fringe induced by a range timing error $\delta\tau_{\mathrm{M},2\pi}$ computed from eq. (3.22). The sensitivity to $B_{\perp}$ outweighs that to $B_{\parallel}$ by one order of magnitude. **(c-d)** Coregistration offset patterns that show a maximum offset of 1 pixel, computed from eq. (3.22) with $f_{\mathrm{PRF}} = 1652\,\mathrm{Hz}$ and $f_{\mathrm{RSR}} = 19.21\,\mathrm{MHz}$.

There is no comprehensive study on the potential magnitude of absolute timing errors that is known to the author. An independent validation of the timing is possible by estimating the shift of the SAR amplitude image with respect to an artificial amplitude image simulated from a DEM (DUT, 2009). M. Arıkan (DUT, pers. comm., 2012) thus observed timing errors of 4 pixels in azimuth and -4 pixels in range, which vary by less than one pixel for a sample of 15 Envisat interferograms. Even though this suggests that relative timing errors are almost negligible, there are occasionally blunders of considerably high order, possibly caused by errors during SAR processing. A. Schenk (KIT, pers. comm., 2012) observed an azimuth timing error of about 0.76 s (1280 pixels) in one particular ERS image, and the author himself identified a range timing error of 0.30 μs (12 pixels) in an Envisat ASAR product.

The further consideration of timing errors will be subdivided into two parts. Initially, their effect on interferograms with a spherical reference surface will be discussed. Later, the case of scenes with significant topographic variations will be addressed.

Effect on a Flat Terrain

As the reference phase does not vary considerably over azimuth, the effect of a timing error in azimuth is very small. For a spherical reference surface with its constant curvature there is no effect at all. Only for an ellipsoidal reference surface, the curvature of which is variable, a phase bias can be observed. Its characteristic basically depends on three parameters: the perpendicular baseline $B_{\perp}$, the geographic latitude Φ and the magnitude δt of the timing error. Figure 3.8a shows the error signal for a baseline of $B_{\perp} = 500$ m and a huge timing error of $\delta t = 1$ s at a latitude of $\Phi = 54°$. The pattern scales almost linearly with $B_{\perp}$ and δt, whereas phase gradient and orientation vary nonlinearly with Φ. The effect is most pronounced at $\Phi = \pm54°$ where the rate of change of the ellipsoid's curvature in the cross-track plane of Envisat is maximum. But as azimuth timing errors are usually much smaller than in this simulation, their effect on the phase can be considered insignificant for a purely ellipsoidal reference surface.

Timing errors in range (see figure 3.4d) have a largely more significant effect on the phase:

$$\delta\varphi = -\delta\phi_{\mathrm{ref}} = -\frac{\partial\phi_{\mathrm{ref}}}{\partial\theta}\,\delta\theta = \frac{4\pi}{\lambda}B_{\perp}\,\delta\theta = \frac{4\pi B_{\perp}}{\lambda R\tan\theta_{\mathrm{inc}}}\,\delta R = \frac{2\pi f B_{\perp}}{R\tan\theta_{\mathrm{inc}}}\,\delta\tau_{\mathrm{M}} \qquad (3.19)$$

with $f = c/\lambda$. Taylor series expansion at the scene centre (t_0, R_0) yields:

$$\delta\varphi(R) = 2\pi f\, \delta\tau_{\mathrm{M}} \left\{ \frac{B_\perp(R_0)}{R_0 \tan\theta_{\mathrm{inc}}(R_0)} \right.$$
$$\left. - \frac{1}{R_0^2} \left[\frac{B_\|(R_0)\, R_0\, \frac{\partial\theta}{\partial R}(R_0) + B_\perp(R_0)}{\tan\theta_{\mathrm{inc}}(R_0)} + \frac{B_\perp(R_0)\, R_0\, \frac{\partial\theta_{\mathrm{inc}}}{\partial R}(R_0)}{\sin^2\theta_{\mathrm{inc}}(R_0)} \right] dR + \dots \right\} \qquad (3.20)$$

with $\partial\theta/\partial R$ and $\partial\theta_{\mathrm{inc}}/\partial R$ as in eqs. (A.1) and (A.2).

The influence of relative timing errors (3.18) on orbital coregistration offsets (2.20) is straightforward:

$$\delta\Delta\breve{\xi} = f_{\mathrm{PRF}} \cdot \delta\Delta t$$
$$\delta\Delta\breve{\eta} = f_{\mathrm{RSR}} \cdot \delta\Delta\tau\,. \qquad (3.21)$$

As it has been done for baseline errors in eq. (3.17), the sensitivity to timing errors can be characterised by the amount of error that is required to induce a maximum phase difference of 2π within one scene or a maximum coregistration offset of 1 pixel, respectively:

$$\delta t_{\mathrm{M},2\pi} \approx 0 \qquad\qquad \delta\Delta t_\xi = f_{\mathrm{PRF}}^{-1} \qquad \delta\Delta t_\eta = 0$$
$$\delta\tau_{\mathrm{M},2\pi} = -\frac{R^2}{f\,\Delta R} \left[\frac{B_\| R\, \frac{\partial\theta}{\partial R} + B_\perp}{\tan\theta_{\mathrm{inc}}} + \frac{B_\perp R\, \frac{\partial\theta_{\mathrm{inc}}}{\partial R}}{\sin^2\theta_{\mathrm{inc}}} \right]^{-1} \qquad \delta\Delta\tau_\xi = 0 \qquad \delta\Delta\tau_\eta = f_{\mathrm{RSR}}^{-1}\,. \qquad (3.22)$$

The corresponding error patterns are displayed in figure 3.8 for Envisat IS2. In contrast to baseline errors, the sensitivity of the phase to range timing errors depends on the absolute baseline, whereof the perpendicular component $B_\perp$ is most determining. The induction of one fringe per 6.7 pixel range timing error for $B_\| = 0\,\mathrm{m}$ and $B_\perp = 500\,\mathrm{m}$ suggests that the influence of timing errors can indeed become significant. On the other hand, timing errors of a fraction of a pixel are identifiable by comparing coregistration offsets from orbits with coregistration offsets from correlation. Thus, they can easily be confined to a level at which their influence on the phase is negligible.

Effect on a Non-Flat Terrain

The above evaluation of timing error effects applies well to interferograms of flat regions but is mostly inadequate in presence of distinct topographic height variations. In this case, the error signal does not only depend on the baseline but also on the local terrain slope. From the illustrations in figure 3.9 can be seen that the effect of azimuth timing errors on the phase is only sensitive to slopes in azimuth, whereas the effect of range timing errors is influenced by slopes in any direction. The simulations apparently suggest that the effect is insignificant in view of the huge timing errors and altitude ranges of 3 km that are required to produce a distinguishable signal. However, locally confined slopes can be considerably steeper when spanning even smaller height differences. The rougher the terrain, the stronger is the local phase bias due to timing errors. On the other hand, a rough terrain makes it easier to estimate and correct for timing errors by exploiting their influence on coregistration offsets. δt and $\delta\tau$ can be estimated from the shift of the master amplitude image with respect to an artificial amplitude image simulated from an accurately georeferenced DEM (DEM-assisted coregistration; ARIKAN ET AL., 2008).

This approach fails for regions with a smooth topography or lacking distinct topographic features, respectively. Whereas the contribution of topography to the phase bias can mostly be neglected in this case, general biases due to range timing errors as simulated in figure 3.8b may still be an issue. Of special concern is their separability from contributions of baseline errors $\delta B_\perp$ (figure 3.5b). But separation should be feasible in a network of interferograms by exploiting either the correlation of timing error signals with $B_\perp$ or even better the sensitivity of coregistration offsets to range timing errors (figure 3.8d).

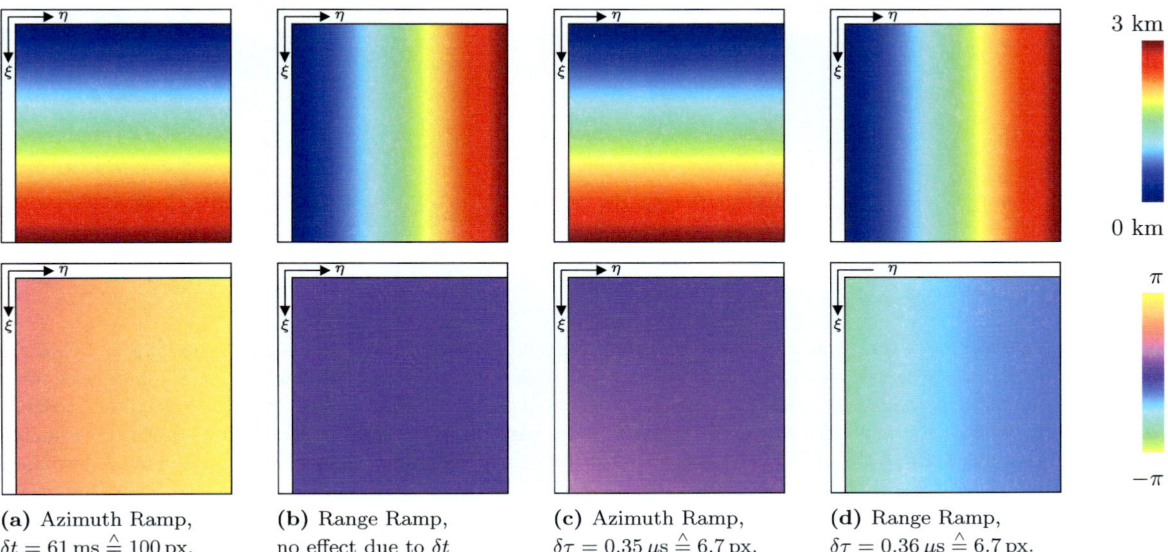

(a) Azimuth Ramp,
$\delta t = 61\,\mathrm{ms} \triangleq 100\,\mathrm{px}$.

(b) Range Ramp,
no effect due to δt

(c) Azimuth Ramp,
$\delta\tau = 0.35\,\mu\mathrm{s} \triangleq 6.7\,\mathrm{px}$.

(d) Range Ramp,
$\delta\tau = 0.36\,\mu\mathrm{s} \triangleq 6.7\,\mathrm{px}$.

Figure 3.9.: *Additional* contribution of topography to the patterns induced by timing errors, computed for Envisat IS2 interferograms with $B_{\parallel} = 0\,\mathrm{m}$ and $B_{\perp} = 500\,\mathrm{m}$ and ramp-like generic DEM. The upper row shows the respective DEM and the lower row the corresponding error signals. The latter scale approximately linearly with the terrain slope and nonlinearly with $B_{\perp}$ and the error itself.

3.4.3. Clock Errors

Strictly speaking, clock errors are not subsumable by orbit errors, because they rather result from an unstable on-board clock than from inaccurate orbit determination. But as clock errors induce linear phase trends that may be mistaken for orbital error signals, it seems appropriate to complementarily discuss this additional error source. Any eventual drift in the sensor's oscillator translates to all derived frequencies, which are besides the carrier frequency f also the pulse repetition frequency f_{PRF} and the range sampling rate f_{RSR}. Assuming the reference oscillator as common error source, the respective frequency errors are linked by the relative error:

$$\nu := \frac{\delta f_{\mathrm{PRF}}}{f_{\mathrm{PRF}}} = \frac{\delta f_{\mathrm{RSR}}}{f_{\mathrm{RSR}}} = \frac{\delta f}{f} \; . \tag{3.23}$$

Effect on the Interferometric Phase

As long as coregistration is implemented by amplitude cross-correlation, the interferometric phase measurement is completely insensitive to errors in f_{PRF} and f_{RSR}, whereas carrier frequency biases δf can indeed produce significant artefacts. These have mostly been beyond the focus of research, because they were neither expected nor are they easy to validate. An early observation was reported by MASSONNET AND VADON (1995) who identified clock errors as the cause of a dubious signal component by excluding all other conceivable error sources. They observed very large short-term carrier frequency drifts up to 82 Hz/s ($\triangleq$ 15 ppb/s) in a long swath ERS-1 interferogram that started and stopped abruptly and lasted some tens of seconds. MASSONNET AND VADON (1995) also describe how relative frequency shifts or drifts between master and slave acquisition translate to error patterns in the interferometric phase. These mechanisms will be recapitulated in the following.

Defining δf_{M} and δf_{S} as the respective deviations of the carrier frequencies from their nominal value f, the relative frequency error is defined as:

$$\delta\Delta f := \delta f_{\mathrm{S}} - \delta f_{\mathrm{M}} \; . \tag{3.24}$$

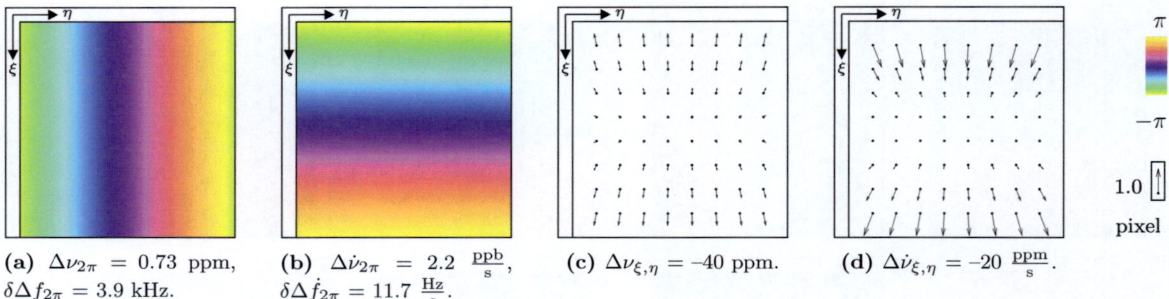

(a) $\Delta\nu_{2\pi} = 0.73$ ppm, $\delta\Delta f_{2\pi} = 3.9$ kHz.   **(b)** $\Delta\dot{\nu}_{2\pi} = 2.2\ \frac{\text{ppb}}{\text{s}}$, $\delta\Delta\dot{f}_{2\pi} = 11.7\ \frac{\text{Hz}}{\text{s}}$.   **(c)** $\Delta\nu_{\xi,\eta} = -40$ ppm.   **(d)** $\Delta\dot{\nu}_{\xi,\eta} = -20\ \frac{\text{ppm}}{\text{s}}$.

Figure 3.10.: Patterns in the phase and the orbital coregistration offsets (2.20), respectively, that are induced by relative frequency offsets and drifts into an Envisat IS2 interferogram with a spherical reference surface. **(a-b)** Error patterns in the interferometric phase showing a maximum phase difference of 2π. The frequency errors that are required to induce these patterns have been computed from eq. (3.27). **(c-d)** Error patterns in the coregistration offsets showing a maximum two-dimensional offset *difference* of 1 pixel. The frequency errors that are required to induce these patterns have been computed from eq. (3.31) with $f_{\text{PRF}} = 1652$ Hz and $f_{\text{RSR}} = 19.21$ MHz.

Its translation into the interferometric phase is described by the following relation:

$$\delta\varphi = -\frac{4\pi}{c}(R_{\text{M,ref}} \cdot \delta f_{\text{M}} - R_{\text{S,ref}} \cdot \delta f_{\text{S}}) \approx \frac{4\pi R}{c}\delta\Delta f \, , \tag{3.25}$$

where $R := R_{\text{M}} \approx R_{\text{M,ref}} \approx R_{\text{S,ref}}$. Taylor series expansion in azimuth time t and range R yields:

$$\delta\varphi(t, R) = \frac{4\pi}{c}\left(R \cdot \delta\Delta f(t_0) + R \cdot \delta\Delta\dot{f}(t_0)\, dt + \delta\Delta f(t_0)\, dR + \dots\right) . \tag{3.26}$$

By analogy to baseline errors and timing errors, the sensitivity of the phase to clock errors is characterised by the amount of error required to induce a maximum phase difference of 2π, given in terms of both absolute and relative errors:

$$\begin{aligned}
\delta\Delta f_{2\pi} &= \frac{c}{2\,\Delta R} & \Delta\nu_{2\pi} &= \frac{\lambda}{2\,\Delta R} \\
\delta\Delta\dot{f}_{2\pi} &= \frac{c}{2R\,\Delta t} & \Delta\dot{\nu}_{2\pi} &= \frac{\lambda}{2R\,\Delta t}
\end{aligned} \tag{3.27}$$

From the corresponding error patterns in figures 3.10a-b can be seen that a relative frequency offset $\delta\Delta f = f \cdot \Delta\nu$ induces fringes in range, whereas a relative frequency drift $\delta\Delta\dot{f} = f \cdot \Delta\dot{\nu}$ causes fringes in azimuth.

Envisat Frequency Decay

A potential occurrence of clock errors has been observed by the scientific community in the form of temporally correlated range trends in time series of Envisat interferograms. KETELAAR (2009, p. 133 et seqq.) documented a very significant trend in PS velocity estimates of $\Delta D' = \partial\Delta D/\partial T = 15$ mm/a over a full scene (100 km ground range, $\Delta R = 39$ km). This trend cannot be caused by orbit errors, because their effects are expected to average out in a time series. It cannot be a deformation signal either, because it is not present in an ERS time series of the same region. However, a long-term drift of the ASAR carrier frequency in the order of $\delta f' = \partial\delta f/\partial T = -2.1$ kHz/a would be a plausible explanation, which is evident from eqs. (3.26), (2.9) and (2.7) or from the following rule of thumb, respectively:

$$\nu' = \frac{\delta f'}{f} = -\frac{\Delta D'}{\Delta R} \, . \tag{3.28}$$

Such a drift would imply frequency offsets between master and slave that are proportional to the temporal baseline:

$$\delta\Delta f = \delta f' \cdot B_T \, . \tag{3.29}$$

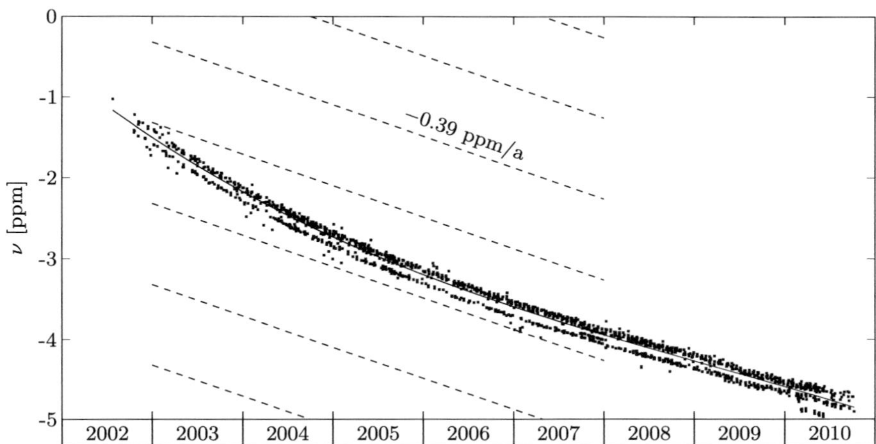

Figure 3.11.: PRF decay estimated from the raw images of about 7000 Envisat scenes by MARINKOVIC AND LARSEN (2013). Their measurements provide evidence for a systematic frequency drift of the ASAR reference oscillator, which would be a plausible explanation for the systematic trend of 15 mm/a per 100 km ground range observed by KETELAAR (2009) in an InSAR time series from 2003 to 2007. This trend could be caused by a drift of the carrier frequency of $\delta f' = -2.1$ kHz/a or $\nu' = \delta f'/f = -0.39$ ppm/a, respectively. (Figure kindly provided by P. Marinkovic, with additions)

From the technical point of view, the occurrence of a frequency drift of this order is definitely within the realms of possibility but would not have violated the mission performance requirements (N. Miranda, ESA, pers. comm., 2012). Further evidence was provided by MARINKOVIC AND LARSEN (2013) who observed a systematic PRF decay in an equivalent order (see figure 3.11). Hence, it seems justified to assume that the whole Envisat mission is affected by a continuous frequency decay on the ASAR instrument.

Effect on Coregistration Offsets

By means of relative biases in the pulse repetition frequency f_{PRF} and the range sampling rate f_{RSR}, clock errors may also translate into orbital coregistration offsets. The effect can be described by Taylor series expansion at the scene centre (t_0, R_0):

$$
\begin{aligned}
\delta\Delta\breve{\xi}(t) &= \delta\Delta\breve{\xi}(t_0) - \delta\Delta f_{\mathrm{PRF}}(t_0)\, dt - \frac{1}{2}\,\delta\Delta\dot{f}_{\mathrm{PRF}}(t_0)\, dt^2 + \ldots \\
\delta\Delta\breve{\eta}(t, R) &= \delta\Delta\breve{\eta}(t_0, R_0) - \frac{2}{c}\left(\delta\Delta f_{\mathrm{RSR}}(t_0)\, dR + \delta\Delta\dot{f}_{\mathrm{RSR}}(t_0)\, dt\, dR + \ldots\right).
\end{aligned}
\tag{3.30}
$$

The zero order coefficients cannot be specified in more detail, because they also comprise secondary effects of clock errors like biased acquisition start times t_1 or biased sampling window start times τ_1. These cannot be characterised in a generic manner. Nevertheless, constant biases of coregistration offsets may also be regarded as timing errors, which have been covered by the previous subsection. Ignoring the constant components restricts the following considerations to offset differences or relative offsets, respectively.

Admitting relative frequency shifts $\Delta\nu = \nu_{\mathrm{S}} - \nu_{\mathrm{M}}$ and drifts $\Delta\dot{\nu} = \dot{\nu}_{\mathrm{S}} - \dot{\nu}_{\mathrm{M}}$, the sensitivity of orbital coregistration offsets with respect to clock errors is characterised by the amount of error that induces a maximum *relative* offset of 1 pixel. As biases in f_{PRF} and f_{RSR} are linked via eq. (3.23), offsets in azimuth and range are considered combinedly in a two-dimensional context:

$$
\begin{aligned}
|\Delta\nu_{\xi/\eta}| &= \max\left\{\frac{1}{f_{\mathrm{PRF}}\,\Delta t},\, \frac{c}{2 f_{\mathrm{RSR}}\,\Delta R}\right\} \\
|\Delta\dot{\nu}_{\xi/\eta}| &= \left(\left(\frac{f_{\mathrm{PRF}}\,\Delta t^2}{8}\right)^2 + \left(\frac{f_{\mathrm{RSR}}\,\Delta t\,\Delta R}{2c}\right)^2\right)^{-\frac{1}{2}}.
\end{aligned}
\tag{3.31}
$$

The corresponding error patterns are displayed in figures 3.10c-d for Envisat IS2. It is evident from the annotations that coregistration offsets are considerably less sensitive to clock errors than the interferometric phase. The identifiability of clock errors by comparison of offsets from orbits and correlation is ultimately limited by a most optimistic accuracy of 1/20 pixel for image correlation (HANSSEN, 2001, p. 46). This would imply a maximum accuracy of 2 ppm or 1 ppm/s, respectively, for the inference of frequency errors from coregistration. However, frequency errors at this level would already induce 2.7 fringes in range or 455 fringes in azimuth, respectively. Against this background it can be concluded that no additional benefit can be drawn from coregistration offsets for the identification of clock errors.

3.5. Conclusions and Outlook

Among the error mechanisms discussed in this chapter, baseline errors have the most disturbing effect on SAR interferograms. Significant error signals are induced either by errors $\delta B_\perp$ in the perpendicular baseline $B_\perp$ or by errors $\delta \dot{B}_\parallel$ in the rate of change of the parallel baseline $B_\parallel$. Timing errors can also have significant effects on the phase. However, these can be confined to an insignificant level in most cases by inferring timing corrections from orbital coregistration offsets. In view of the evidence of a systematic frequency decay on the Envisat ASAR instrument, potentially occurring clock errors and their resulting interferometric error signals are definitely worth considering.

The accuracy of orbit products is very heterogeneous for the currently operational SAR satellites and depends substantially on the priority of InSAR applications for the satellite design. Recent advances in orbit determination are mainly due to the availability of better gravity models and new measurement techniques with good global coverage. The solar minimum in late 2008 additionally promoted some excellent POD results in the last decade.

For most interferometric pairs from most spaceborne SAR sensors, baseline errors can be expected to be in the order of 1 dm or below, occasionally ranging up to several decimetres. Only for Radarsat-1/2, errors on the decimetre or even metre level are rather the rule than the exception. An orbit accuracy about 2 cm, which has already been proven achievable for TerraSAR-X, is considered expectable for future missions. Eventually, the inclusion of complementary GNSS besides the GPS can further enhance reliability and thus reduce the outlier rate.

As the relevance of InSAR deformation analysis is increasingly recognised, the design of new missions more often accommodates the requirement of short baselines. Thus, orbital tubes of only 100 m width are envisaged for both Sentinel-1 (GEUDTNER ET AL., 2011) and the Radarsat constellation (THOMPSON, 2010). Geometric decorrelation due to convergent orbits will probably be entirely negligible under these conditions. However, a tighter orbit control also requires more frequent maneuvers, which increases again the challenges of orbit determination.

4. Baseline Error Estimation

After the inference of interferometric phase patterns from given baseline errors has been discussed in section 3.4.1, this is dedicated to the inverse problem, namely inferring baseline error parameters from the residual phase pattern of a single interferogram. Basic aspects of methodology and estimability will be discussed in the context of existing approaches before two estimators are proposed in section 4.3. Finally, aspects of parameterisation and reliability will be subject to an elaborate evaluation and discussion.

All approaches for baseline error estimation addressed in this chapter assume that except for orbital error signals $\delta\phi_{\mathrm{orb}}$, all contributions to the residual interferometric phase φ in eq. (2.7) behave randomly. This assumption implies that atmospheric and deformation signals leak into the baseline estimates if they are not removed beforehand. As removal is usually not straightforward to achieve, resistance with respect to unmodelled non-orbital signals will be considered a central requirement for the optimal design of an estimator. Capabilities and limitations of the stochastic model to account for non-orbital contributions will be discussed in section 4.6. Separability of signal components will additionally be addressed in section 6.4.5.

The latter remark already adumbrates that chapter 4 is indeed embedded into a greater context spanning three chapters. The concepts of baseline error estimation from individual interferograms discussed in this chapter are extended to simultaneous orbit error estimation from multiple interferometric combinations in chapter 5. Finally, the methodology is tested and evaluated in chapter 6 by application to a real data set. In compliance with this three-part structure, some relevant conclusions are not drawn before the end of chapter 6.

4.1. Estimability

Before reviewing existing estimation approaches, estimability of baseline errors is analysed from the geometric point of view. Given an orbital error signal $\delta\phi_{\mathrm{orb}}$ in the interferometric phase, baseline error estimation is equivalent to inferring the slave orbit if the master orbit is deterministically known. Initially, the problem is considered in the two-dimensional cross-track plane. As the phase is ambiguous, the absolute phase measurement is arbitrary, but the phase difference between two pixels 1 and 2 can be directly related to a double difference of ranges:

$$\Delta\phi_{12} = -\frac{4\pi}{\lambda}\,\Delta R_{12} = -\frac{4\pi}{\lambda}\big((R_{\mathrm{M},2} - R_{\mathrm{S},2} - (R_{\mathrm{M},1} - R_{\mathrm{S},1})\big)\,. \tag{4.1}$$

With a fixed master position M, all potential slave positions S that have a constant phase difference with respect to two points 1 and 2 on the surface lie on a hyperbolic position line (see figure 4.1a) bisecting the parallactic angle between 1 and 2 at S. Hence, the location of S can be inferred in a direction perpendicular to the line of sight from one phase difference. Whereas the perpendicular baseline $B_{\perp}$ can be identified with the distance of M from this position line, it is impossible to obtain the full baseline from one phase difference only. If, however, two phase differences are measured (see figure 4.1b), the

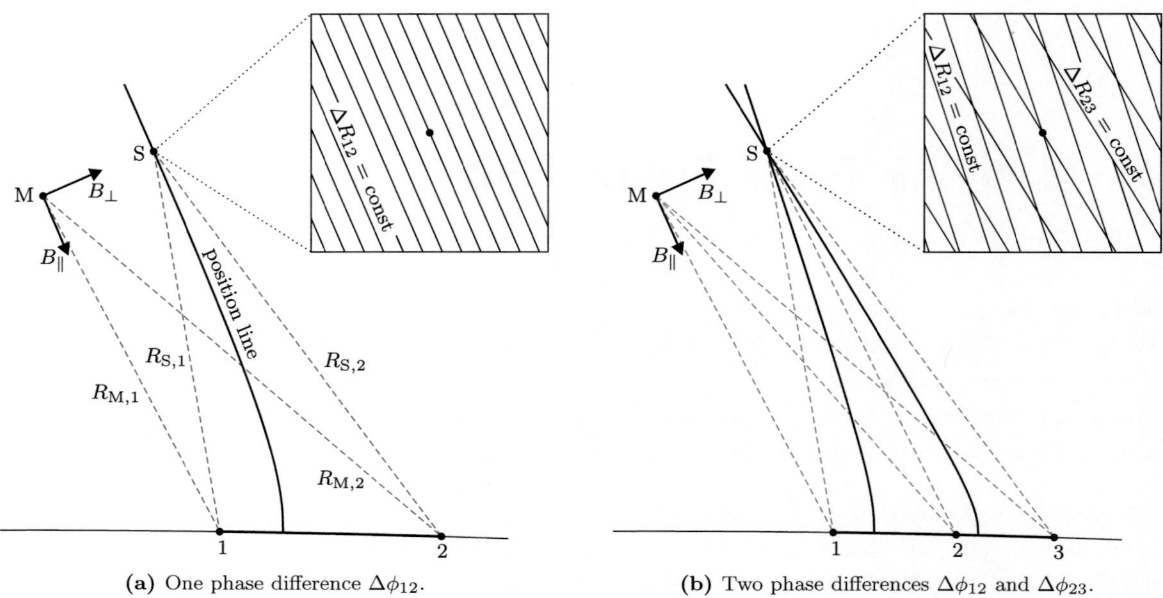

(a) One phase difference $\Delta\phi_{12}$. **(b)** Two phase differences $\Delta\phi_{12}$ and $\Delta\phi_{23}$.

Figure 4.1.: Inference of the slave position S from interferometric phase differences, assuming a deterministic master position M. The location of S can be determined most precisely perpendicular to the LOS, whereas its component in the LOS-direction is not estimable from one phase difference and still poorly determined from two phase differences, respectively. The indicated baseline components $B_\parallel$ and $B_\perp$ are aligned here with the directions of minimum and maximum estimation quality, respectively.

corresponding position lines have a unique intersection point, enabling the determination of the full two-dimensional baseline. But due to the glancing intersection, only $B_\perp$ is estimable with a high precision, whereas $B_\parallel$ is poorly determined.

It is also obvious from figure 4.1 that the density of the position lines and thus the achievable precision of the estimated position S increases with the spatial separation of the two ground points. The estimation quality can further be enhanced by using more than two phase differences and obtaining an overdetermined solution. Alternatively, absolute phases can be used instead of differences if a constant phase offset parameter is jointly estimated.

According to the baseline definition in section 3.1, the decomposition into $B_\parallel$ and $B_\perp$ is not unique and depends on the respective line of sight to a specific pixel. Hence, it is not rigorously accurate to state that "$B_\perp$" can be estimated from a phase difference between *two* pixels, each of which is implying an individual baseline decomposition. This ambiguity is overcome by the following convention, which is convenient in the context of baseline error estimation: $B_\perp$ denotes always the baseline component in the direction of highest estimation quality, and $B_\parallel$ represents the component that is estimable with the lowest precision.

The characterisation of the azimuth component of orbital error signals in the interferometric phase is relatively simple compared to the range component. The most significant influence has a variation of the baseline error component $\delta B_\parallel(t)$ in LOS direction, since it translates directly, both quantitatively and qualitatively, into a phase variation in azimuth (see figure 4.2):

$$\delta\phi_{\mathrm{orb}}(t) = -\delta\phi_{\mathrm{ref}}(t) \approx \frac{4\pi}{\lambda}\,\delta B_\parallel(t)\,. \tag{4.2}$$

Whereas the estimation of variations in $\delta B_\parallel$ is straightforward, variations in $\delta B_\perp(t)$ are also estimable from the theoretical point of view, because they cause a change of the error signal in range over azimuth. But this is only a second order effect, and the influence on the phase is less dominant.

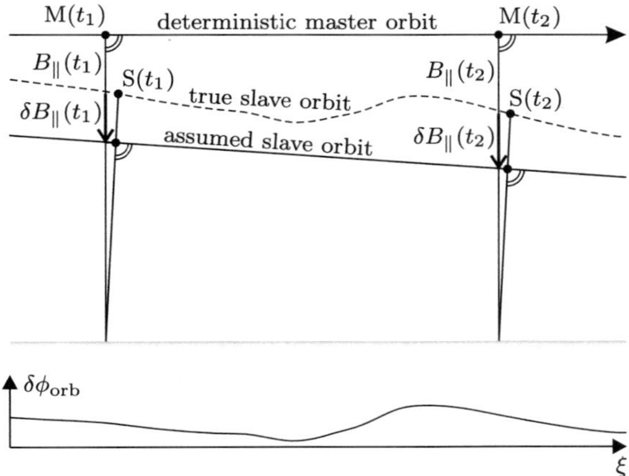

Figure 4.2.: Temporally varying error $\delta B_\|(t)$ in the parallel baseline, projected onto a plane defined by the lines of sight of the master acquisition. According to eq. (4.2), the error translates directly into a phase bias $\delta\phi_\mathrm{orb}(t)$.

4.2. Classification of Approaches

A number of approaches have already been proposed for baseline error estimation that are all similar or different in a number of respects. As a general classification would not account for all individual characteristics, an adequate overview is provided by classifying different methodological aspects. Whereas the first approaches have been developed in the context of topographic mapping in the early 1990s, the following discussion focusses on deformation monitoring as intended application.

4.2.1. Relevance of Auxiliary Data

Given precise orbits and a DEM, the interferometric baseline $\vec{B}(t, R)$ (and thus the reference phase) can be computed for every individual pixel as a function of azimuth time t and range R. Due to the finite accuracy of orbit products, some authors refer to this procedure as the "orbit method" of baseline estimation (ZHANG ET AL., 2009; ZHAO ET AL., 2010). However, it is more common to consider it a mere computation of an initial baseline, which can be subject to further refinement by estimation of a residual baseline *error* $\delta\vec{B}$. In terms of processing this means that the reference phase is computed from the orbits and subtracted from the interferogram in order to estimate a residual baseline error from the residual phase.

If no orbit data are available, baseline estimation from an interferogram is still possible based on a very approximate master orbit. The lack of a priori information requires a relatively fine parameterisation to adequately characterise the baseline vector and its time evolution during the acquisition. However, admitting many degrees of freedom implies an increased risk that unmodelled nuisance signals in the phase leak into the baseline estimates. This risk is considerably mitigated by estimating only a small update to an initial baseline computed from orbit data, since a residual baseline error can be described with sufficient accuracy by a small number of parameters.

In case of unknown topography, it is possible to yield a less accurate baseline estimate by assuming the bare ellipsoid as reference surface. Whereas this approach is acceptable for flat regions, it can cause seriously biased baseline estimates in presence of steep topography.

4.2.2. Distinction from Deformation

To avoid or mitigate the leakage of deformation signals into the baseline error estimates, one of the following approaches can be pursued:

1. *Explicit ground truth.* Estimation is performed at a number of selected ground control points, the displacement of which is known for instance from GNSS measurements (e. g., ZEBKER ET AL., 1994a, p. 19630; MURAKAMI ET AL., 1996; LUNDGREN ET AL., 2009). At these points, the contribution of deformation is a priori subtracted from the phase.

2. *Implicit ground truth.* Estimation is based on the assumption that there is no ground deformation in a particular subregion (e. g., ELLIOT ET AL., 2008), which is mostly not straightforward to validate. The approach is only applicable if the extent of the interferogram is larger than the region that is subject to the deformation.

3. *Iteratively alternating estimation* of orbital contribution and deformation in dedicated parametric models (BIGGS ET AL., 2007). This approach does not preserve long wavelength deformation components.

4. *Spatial filtering.* Assuming a maximum wavelength for deformation signals, contributions below a threshold wavelength can be filtered out, for instance by means of a wavelet decomposition (SHIRZAEI AND WALTER, 2011). Obviously, this approach does not preserve long wavelength deformation components either.

5. *Temporal filtering.* A posteriori high pass filtering of the orbit error estimates in time can remove temporally correlated contributions of deformation that have erroneously been attributed to the orbital component (BÄHR AND HANSSEN, 2012). Conversely, the estimated deformation signal can be low pass filtered in time to remove residual orbital contributions. The latter approach is the preferred one if deformation rather than orbital effects is the signal of interest. It does not require explicit baseline error estimation provided that orbit errors are not extraordinarily large. It is furthermore (either explicitly or implicitly) pursued by all PS processors.

Besides deformation, also atmospheric contributions can leak into the baseline estimates. However, as long as deformation is the signal of interest, this leakage can be tolerated. Analogously, leakage of large-scale deformation signals into the orbital component is acceptable when analysing the (short scale) atmospheric contribution.

4.2.3. Observations

Baseline estimation techniques are based on one of the following observation types:

1. *Unwrapped phase differences* can be obtained from counting fringes along the edges of an interferogram (MASSONNET AND FEIGL, 1998, p. 458 et seq.; KOHLHASE ET AL., 2003). PEPE ET AL. (2011) rely on phase differences between adjacent pixels, which makes the estimates insensitive to spatial propagation of unwrapping errors at the expense of a reduced information content per observation.

2. *Unwrapped phases* can alternatively be used if a global phase offset is introduced as an additional parameter (SMALL ET AL., 1993; WERNER ET AL., 1993; ZEBKER ET AL., 1994b; BIGGS ET AL., 2007; ZHANG ET AL., 2009; SHIRZAEI AND WALTER, 2011; BÄHR AND HANSSEN, 2012).

3. *Wrapped phases* can be exploited by estimators based on an integral transformation of the interferometric signal (SINGH ET AL., 1997; MONTI GUARNIERI ET AL., 2000; BING ET AL., 2006; BÄHR AND HANSSEN, 2012).

4. *Orbital coregistration offsets in range* can provide baseline estimates even in case of lacking information on precise orbits and terrain, but with inferior accuracy. KNEDLIK ET AL. (1999) obtain residual baseline errors of several decimetres and larger from simulations with an optimistic assumption of 0.03 pixels for the coregistration accuracy.

Estimators operating in the spatial domain are based on either unwrapped phase differences (1.) or unwrapped phases (2.). The second option is considered the more generic one, although it involves the introduction of an additional offset parameter. Rigorously exploiting phase differences turns out to be more complicated, because it requires the definition of pixel pairings. If n unwrapped phase observations are paired to form $n/2$ phase differences, not all available information is used. If however $n - 1$ phase differences are formed from n unwrapped phase observations, the phase differences are not stochastically independent. The resulting correlations between observations can be considered in the stochastic model, but this would increase both the computational and the conceptional load significantly.

4.2.4. Selection of Observations

Among all pixels of an interferogram, phase observations can be selected by one of the following strategies:

1. *Interferogram corners.* Counting fringes along the edges of an interferogram yields three linearly independent phase differences with most optimal spatial coverage (MASSONNET AND FEIGL, 1998, p. 458 et seq.; KOHLHASE ET AL., 2003).

2. *Ground control points (GCP).* Individual pixels with known deformation are selected (ZEBKER ET AL., 1994a, p. 19630; MURAKAMI ET AL., 1996; LUNDGREN ET AL., 2009). For baseline estimation in the context of topographic mapping this may also be pixels with known topographic height (SMALL ET AL., 1993; PRATI ET AL., 1993; ZEBKER ET AL., 1994b).

3. *All pixels.* Using all available pixels can only yield suitable results if the expectable outlier rate due to decorrelation is low, for instance in ERS tandem interferograms (MONTI GUARNIERI ET AL., 2000), single-pass tandems (BING ET AL., 2006) or coherent subframes (SINGH ET AL., 1997).

4. *Selection by coherence.* Only pixels with a coherence above a dedicated threshold are used (BIGGS ET AL., 2007). SHIRZAEI AND WALTER (2011) use all pixels but apply a coherence-based weighting in an initial iteration.

5. *Homogeneously distributed subset.* Defining a regular grid on the interferogram, the most coherent pixel is selected from every grid cell (BÄHR AND HANSSEN, 2012).

It should be noted that the estimates can be very sensitive to spatially inhomogeneous distributions of observations or inhomogeneously distributed weights thereof. These may cause so-called leverage effects, which will be discussed in detail in section 4.5.1.

4.2.5. Estimator

Estimation strategies can be classified into the following types:

1. *Consistently determined solution.* Due to lack of redundancy, this class does not represent estimators in its strict sense. For instance, (ZEBKER ET AL., 1994b) solve for a parameter set from an equal number of observations; MASSONNET AND FEIGL (1998, p. 458 et seq.) adjust three residual phase differences to zero by modifying the orbital trajectories in three degrees of freedom. Consistently determined solutions generally require unwrapping.

2. *Least Squares Adjustment.* Solving an overdetermined equation system by minimising the sum of squared phase residuals is a very popular approach that usually requires unwrapping (SMALL ET AL., 1993; WERNER ET AL., 1993; BIGGS ET AL., 2007; SHIRZAEI AND WALTER, 2011; BÄHR AND HANSSEN, 2012). The implementation of BÄHR AND HANSSEN (2012) is presented in detail in section 4.3.1.

3. *Periodogram-based estimation.* Fourier transformation of the wrapped interferogram enables exploiting the direct relation (2.22) between the baseline component perpendicular to the line of sight and the local fringe frequency in range (PRATI AND ROCCA, 1990). Theoretically, the full two-dimensional baseline error $(\delta B_h, \delta B_v)$ can be inferred from the one-dimensional periodogram in range (SINGH ET AL., 1997) by exploiting both the centre frequency and the bandwidth of the dominant peak in the data spectrum. The component $\delta B_\parallel$ in look direction is only weakly determined though, as it can be expected from the estimability considerations in section 4.1. The direct estimation of one well estimable parameter $\delta B_\perp(R_0)$ from the peak frequency using eq. (2.22) (BING ET AL., 2006) is minimally biased, since the peak in the spectrum comprises the dominant fringe frequencies from all ranges while serving as an estimate for the local fringe rate at R_0. Thus, the nonlinear characteristics of the orbital error signal in range are negated. Inference of $\delta \dot{B}_\parallel$ from the dominant fringe frequency in azimuth is straightforward and can be performed independently from the range component.

4. *Integral transform to the $(\dot{B}_\parallel, B_\perp)$-space.* This approach is similar to the periodogram-based estimation but is unbiased due to rigorous geometric modelling. A gridsearch implementation (BÄHR AND HANSSEN, 2012) is discussed in detail in section 4.3.2.

5. *Slope PDF.* For the case of lacking precise orbits and unknown but significant topography, MONTI GUARNIERI ET AL. (2000) propose to estimate $B_\perp$ by adjusting a parametric model to the empirical probability density function (PDF) of terrain slopes. Whereas the approach outperforms the periodogram-based estimator for dedicated applications, it yields no improvement if a DEM is available.

6. *Least mode of binned residuals.* PEPE ET AL. (2011) pursue a heuristic approach, estimating two baseline parameters separately. They obtain one candidate estimate per observation by solving for a parameter from one phase gradient observation at a time. Among all these candidates, the value is selected that minimises the peak location in the histogram of residuals.

ZHAO ET AL. (2010) propose a sequential application of two estimators. However, the observed improvement can be attributed to the finer parameterisation of the second one and is probably not due to the organisation in two steps.

4.2.6. Parametric Model

The choice of an appropriate parametric model involves a critical trade-off between approximation errors and overparameterisation. Too few parameters may not capture the signal appropriately, and too many parameters allow for leakage from other contributions. This conflict will be discussed in detail in section 4.4.

Baseline error parameterisations are based on one of the following concepts:

1. *Polynomial approximation.* Not the baseline error itself but rather its effect on the phase is parameterised by a polynomial expansion in pixel coordinates (ξ, η) of degree d:

$$p_d(\xi, \eta) = \sum_{i=0}^{d} \sum_{j=0}^{d-i} a_{ij} \left(\frac{\xi - \xi_0}{\Delta \xi} \right)^i \left(\frac{\eta - \eta_0}{\Delta \eta} \right)^j , \qquad (4.3)$$

   whereas the translation by (ξ_0, η_0) and the scaling by $(\Delta \xi, \Delta \eta)$ are arbitrary and made for purposes of numerical stability. a_{00} can be considered a nuisance parameter, because a global offset has no implication for the interpretation of an interferogram. Very common is a linear polynomial $p_1(\xi, \eta)$ or phase ramp, respectively (e. g., ZEBKER ET AL., 1994a, p. 19630; FEIGL ET AL., 2002; BIGGS ET AL., 2007; SHIRZAEI AND WALTER, 2011). This parameterisation is easy to implement but involves a minor bias, since the orbital error signal is not rigorously linear. Whereas quadratic polynomials qualify by a better fit as evaluated by HANSSEN (2001, p. 126 et seq.), they involve the risk of overparameterisation. LUNDGREN ET AL. (2009) adjust the polynomial not to the phase but rather to the assumed LOS displacement rate D, which is mathematically equivalent. Developing the polynomial in look angle θ instead of range R or η, respectively (BÄHR AND HANSSEN, 2012), is numerically equivalent to a parameterisation by $\delta B_\perp$ as will be shown in section 4.4.1.

2. *Baseline parameterisation.* The parameter set consists of selected baseline error components. Solving for $\delta B_\perp$ only (MONTI GUARNIERI ET AL., 2000; BING ET AL., 2006) does not account for variations over azimuth. Neither does the $(\delta B_h, \delta B_v)$-representation (SINGH ET AL., 1997), which implicitly comprises the weakly-determined $\delta B_\parallel$-component. A most stable parameterisation is $(\delta \dot{B}_\parallel, \delta B_\perp)$ (BÄHR AND HANSSEN, 2010, 2012; PEPE ET AL., 2011). Similar results may be obtained with the parameter set $(\delta B_h, \delta \dot{B}_h)$ (SMALL ET AL., 1993; WERNER ET AL., 1993), which basically spans the same solution space but does not coincide with its principal directions (see figure 4.3). ZHANG ET AL. (2009) suggest a strongly overparameterised model of 18 parameters, estimating quadratic polynomials for $\delta B_h(\xi, \eta)$, $\delta B_a(\xi, \eta)$ and $\delta B_v(\xi, \eta)$. The set of 6 parameters $(\delta H_\mathrm{M}, \delta \dot{H}_\mathrm{M}, \delta B_h, \delta \dot{B}_h, \delta B_\perp, \delta \dot{B}_\perp)$ proposed by ZHAO ET AL. (2010) is also questionable, because the phase is almost insensitive to the master sensor height H_M, and the parameter-subspaces of δB_h and $\delta B_\perp$ are not orthogonal.

3. *Mixed parameterisation.* ZEBKER ET AL. (1994b) use a mixed parameterisation of a baseline component and a phase polynomial coefficient from eq. (4.3): $(\delta B_h, a_{10})$. a_{10} can be considered almost equivalent to $\delta \dot{B}_\parallel$ though.

4. *State vector parameterisation.* Instead of interferogram-specific baseline errors $\delta \vec{B}$, also acquisition-specific orbit errors $\delta \vec{x}$ can be inferred, but only in the context of a network of multiple interferograms (see chapter 5). KOHLHASE ET AL. (2003) were probably the first authors to propose this approach with four parameters $(x_h(t_1), x_v(t_1), x_h(t_1 + \Delta t), x_v(t_1 + \Delta t))$ per acquisition. BÄHR AND HANSSEN (2012) used the more stable parameterisation $(\dot{x}_\parallel, x_\perp)$ but proposed a sequential approach, in which baseline errors $(\delta \dot{B}_\parallel, \delta B_\perp)$ are estimated in a first step, and the corresponding orbit errors $(\delta \dot{x}_\parallel, \delta x_\perp)$ are subsequently adjusted to the baseline errors. This stepwise procedure involves a negligible bias, which is discussed in section 5.2.1.

4.2.7. Robustness

A principal requirement for robust estimates are redundant observations. Whereas redundancy itself already involves some degree of resistance with respect to outliers, the resistance can be further enhanced by dedicated robust estimation techniques. These generally confine the influence of individual outliers on the final estimates. For the application of robust techniques, two levels can be distinguished:

1. *Pixel level.* SHIRZAEI AND WALTER (2011) use a dedicated reweighting scheme that iteratively downweights contributions of pixels with outlying phase values. BÄHR AND HANSSEN (2012) propose to apply classical data snooping (BAARDA, 1968), rejecting one outlying observation per iteration completely (see section 4.5.2).

2. *Interferogram level.* Robustness of the estimates can also be enhanced by the inclusion of redundant interferograms in a network of linearly dependent interferometric combinations (see chapter 5; KOHLHASE ET AL., 2003; BIGGS ET AL., 2007; PEPE ET AL., 2011). A systematic rejection scheme of interferograms with outlying baseline estimates has been proposed by BÄHR AND HANSSEN (2012) and will be revisited in detail in section 5.3.

4.3. Mathematical Model

Two estimators for the baseline error parameters $(\delta \dot{B}_\parallel, \delta B_\perp)$ are receiving special attention within the scope of this thesis. Their underlying functional model is closely related to the approach of KOHLHASE ET AL. (2003), differing in some aspects though. The model is derived within the framework of a least squares estimator (BÄHR AND HANSSEN, 2010, 2012) and relies on unwrapped phase observations. It is subsequently adopted for a *gridsearch* estimator (BÄHR AND HANSSEN, 2012) that is based on an integral transform of the wrapped interferogram. Note that the sign convention applied to the observation equations here is different from previous publications (BÄHR AND HANSSEN, 2010, 2012) but consistent with previous chapters of this thesis.

4.3.1. Least Squares Estimator

Orbit errors translate into the interferogram via biased ranges between sensor and target that are assumed for the computation of the reference phase in eq. (2.6). Biases in $\delta R_{M,ref}$ and $\delta R_{S,ref}$ generate a reference phase bias $\delta\phi_{ref}$ that equals the orbital error signal $\delta\phi_{orb}$ after sign inversion. Neglecting all other contributions to the residual phase φ, the following observation equation holds:

$$\mathrm{E}\{\varphi\} = \delta\phi_{orb} = -\delta\phi_{ref} = \frac{4\pi}{\lambda}(\delta R_{M,ref} - \delta R_{S,ref}) + \varphi_0 \,. \tag{4.4}$$

φ_0 is a constant phase offset that accounts for the relative nature of phase measurements. Expressing the range biases $\delta R_{M,ref}$ and $\delta R_{S,ref}$ by a superposition of biases in the horizontal, along-track and vertical components of the assumed orbit positions $\vec{x}_M$ and $\vec{x}_S$, respectively, yields:

$$
\begin{aligned}
\mathrm{E}\{\varphi\} = {} & \frac{4\pi}{\lambda}(\delta R_{M,ref} - \delta R_{S,ref}) + \varphi_0 \\
= {} & \frac{4\pi}{\lambda}\left\{ \frac{\partial R_{M,ref}}{\partial x_{h,M}}\delta x_{h,M} + \frac{\partial R_{M,ref}}{\partial x_{a,M}}\delta x_{a,M} + \frac{\partial R_{M,ref}}{\partial x_{v,M}}\delta x_{v,M} \right. \\
& \left. - \frac{\partial R_{S,ref}}{\partial x_{h,S}}\delta x_{h,S} - \frac{\partial R_{S,ref}}{\partial x_{a,S}}\delta x_{a,S} - \frac{\partial R_{S,ref}}{\partial x_{v,S}}\delta x_{v,S} \right\} + \varphi_0
\end{aligned}
$$

$$
\begin{aligned}
&= \frac{4\pi}{\lambda} \Big\{ \langle \vec{r}_{\mathrm{M,ref}}, \vec{e}_h \rangle \, \delta x_{h,\mathrm{M}} + \langle \vec{r}_{\mathrm{M,ref}}, \vec{e}_a \rangle \, \delta x_{a,\mathrm{M}} + \langle \vec{r}_{\mathrm{M,ref}}, \vec{e}_v \rangle \, \delta x_{v,\mathrm{M}} \\
&\qquad - \langle \vec{r}_{\mathrm{S,ref}}, \vec{e}_h \rangle \, \delta x_{h,\mathrm{S}} - \langle \vec{r}_{\mathrm{S,ref}}, \vec{e}_a \rangle \, \delta x_{a,\mathrm{S}} - \langle \vec{r}_{\mathrm{S,ref}}, \vec{e}_v \rangle \, \delta x_{v,\mathrm{S}} \Big\} + \varphi_0 \\
&=: \; - a_{h,\mathrm{M}} \, \delta x_{h,\mathrm{M}} - a_{a,\mathrm{M}} \, \delta x_{a,\mathrm{M}} - a_{v,\mathrm{M}} \, \delta x_{v,\mathrm{M}} \\
&\qquad + a_{h,\mathrm{S}} \, \delta x_{h,\mathrm{S}} + a_{a,\mathrm{S}} \, \delta x_{a,\mathrm{S}} + a_{v,\mathrm{S}} \, \delta x_{v,\mathrm{S}} + \varphi_0 \, ,
\end{aligned}
\tag{4.5}
$$

where $\vec{r}_{\mathrm{M,ref}} := \vec{r}_M(t, R_{\mathrm{M,ref}})$ and $\vec{r}_{\mathrm{S,ref}} := \vec{r}_S(t, R_{\mathrm{S,ref}})$ are unit vectors describing the assumed line of sight (see figure 3.1b). $\vec{e}_h(t)$, $\vec{e}_a(t)$ and $\vec{e}_v(t)$ form a Frenet frame as defined in eq. (3.2). From the virtual acquisition geometry for zero-Doppler focussed data (see figure 2.1b) follows $a_{a,\mathrm{M}} = \langle \vec{r}_{\mathrm{M,ref}}, \vec{e}_a \rangle = 0$, and also $a_{a,\mathrm{S}} = \langle \vec{r}_{\mathrm{S,ref}}, \vec{e}_a \rangle \approx 0$ holds due to the high degree of orbit collinearity in spaceborne SAR. Hence, the interferometric phase is not sensitive to orbit errors in along-track direction, and their contributions are neglected in the following. Instead, variations of orbit errors in time are allowed for by introducing polynomials of degree d:

$$
\begin{aligned}
\mathrm{E}\{\varphi\} = &- \sum_{i=0}^{d} \left(a_{h,\mathrm{M}} \, t^i \right) \cdot \delta x_{h,\mathrm{M}} - \sum_{i=0}^{d} \left(a_{v,\mathrm{M}} \, t^i \right) \cdot \delta x_{v,\mathrm{M}} \\
&+ \sum_{i=0}^{d} \left(a_{h,\mathrm{S}} \, t^i \right) \cdot \delta x_{h,\mathrm{S}} + \sum_{i=0}^{d} \left(a_{v,\mathrm{S}} \, t^i \right) \cdot \delta x_{v,\mathrm{S}} + \varphi_0 \, .
\end{aligned}
\tag{4.6}
$$

This most general observation equation has $4(d+1)+1$ parameters. But as orbit trajectories are very smooth curves, errors in their determination can adequately be described with a polynomial of low degree. In the following, $d = 1$ will be assumed, since a linearly varying baseline error is considered an appropriate approximation for most applications (see also sections 4.4.2 and 4.4.4).

Moreover, the coefficients in eq. (4.6) are almost identical for master and slave due to the very small divergence between $\vec{r}_M$ and $\vec{r}_S$. This makes the joint estimation of individual orbit errors for both master and slave an ill-posed problem that can only be solved in a network of interferograms (see chapter 5). Considering one interferogram on its own, only components of a relative error $\delta \vec{B} = \delta \vec{x}_S - \delta \vec{x}_M$ can be robustly estimated. In this case, it must be decided if the estimated error is heuristically attributed to inaccuracies in the master orbit, the slave orbit or to errors in both of them. In the following relation, the error is attributed in equal proportions to master and slave in order to avoid an arbitrary discrimination of one of the two acquisitions:

$$
\begin{aligned}
\mathrm{E}\{\varphi\} &= \frac{a_{h,\mathrm{M}} + a_{h,\mathrm{S}}}{2} \left(\delta B_h + t \, \delta \dot{B}_h \right) + \frac{a_{v,\mathrm{M}} + a_{v,\mathrm{S}}}{2} \left(\delta B_v + t \, \delta \dot{B}_v \right) + \varphi_0 \\
&=: a_h \left(\delta B_h + t \, \delta \dot{B}_h \right) + a_v \left(\delta B_v + t \, \delta \dot{B}_v \right) + \varphi_0 \, .
\end{aligned}
\tag{4.7}
$$

Considering the residual interferometric phases $\boldsymbol{\varphi}^T = (\cdots \; \varphi_i \; \cdots)$ of n_φ pixels that are regularly arranged on a grid spanning the whole interferogram, baseline error parameters $\mathbf{b}^T = (\, \delta B_h \; \delta \dot{B}_h \; \delta B_v \; \delta \dot{B}_v \,)$ can be estimated in a functional model of the following kind:

$$
\mathrm{E}\{\boldsymbol{\varphi}\} =
\begin{pmatrix}
\vdots & \vdots & \vdots & \vdots \\
a_{h,i} & a_{h,i} t_i & a_{v,i} & a_{v,i} t_i \\
\vdots & \vdots & \vdots & \vdots
\end{pmatrix}
\begin{pmatrix}
\delta B_h \\
\delta \dot{B}_h \\
\delta B_v \\
\delta \dot{B}_v
\end{pmatrix}
+
\begin{pmatrix}
\varphi_0 \\
\vdots \\
\varphi_0
\end{pmatrix}
=: \mathbf{A}_b \mathbf{b} + \mathbf{1}\, \varphi_0 \, ,
\tag{4.8}
$$

where $\mathbf{1}^T = (\, 1 \; 1 \; \cdots \; 1 \,)$. The stochastic model is generically defined by some covariance matrix:

$$
\mathrm{D}\{\boldsymbol{\varphi}\} = \sigma_0^2 \mathbf{Q}_\varphi \, ,
\tag{4.9}
$$

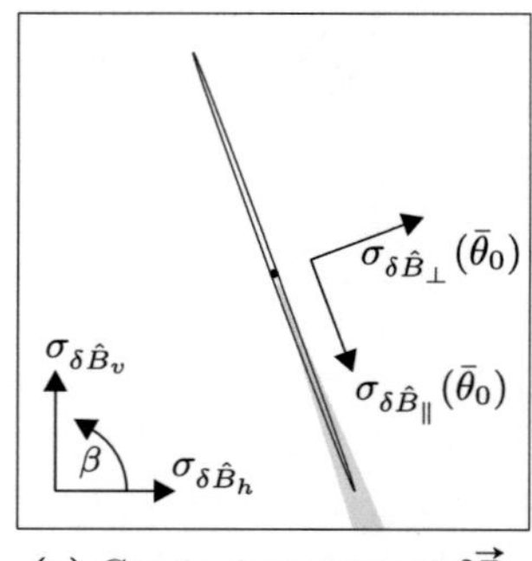

(a) Constant component $\delta \vec{B}$.

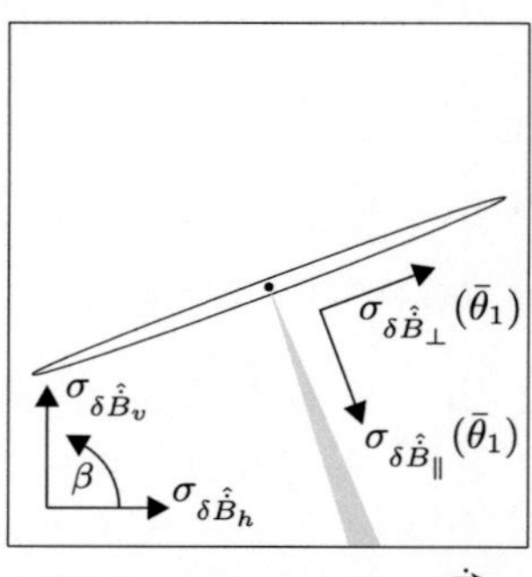

(b) Linear component $\delta \dot{\vec{B}}$.

Figure 4.3.: Anisotropic estimation quality of the baseline error, visualised by error ellipses. The grey area represents the sensor's field of view. The orientation angles of the ellipses with respect to the nadir ($\bar{\theta}_0$ and $\bar{\theta}_1$, respectively) can be computed from the eigenspaces of the covariance matrix $D\{\hat{\mathbf{b}}\}$. It follows that the estimability of $\delta B_\perp(\bar{\theta}_0)$ and $\delta \dot{B}_\parallel(\bar{\theta}_1)$ is best, whereas $\delta B_\parallel(\bar{\theta}_0)$ and $\delta \dot{B}_\perp(\bar{\theta}_1)$ are most weakly determined.

where $D\{\cdot\}$ denotes the dispersion. The choice of $D\{\varphi\}$ is discussed in more detail in section 4.6. As φ_0 is of no further interest, this parameter can be eliminated from eq. (4.8) yielding (TEUNISSEN, 2000, p. 91 et seqq.; NIEMEIER, 2008, p. 307 et seqq.; JÄGER ET AL., 2006, p. 37 et seqq.):

$$E\{\varphi\} = \bar{\mathbf{A}}_b \mathbf{b} \tag{4.10}$$

with:

$$\bar{\mathbf{A}}_b = \left(\mathbf{I} - \mathbf{1}(\mathbf{1}^T\mathbf{Q}_\varphi^{-1}\mathbf{1})^{-1}\mathbf{1}^T\mathbf{Q}_\varphi^{-1}\right)\mathbf{A}_b , \tag{4.11}$$

where $\mathbf{I}$ is the identity matrix.

The relative estimation quality of the least squares estimates of $\hat{\mathbf{b}}$ is given by their covariance matrix:

$$D\{\hat{\mathbf{b}}\} = \sigma_0^2(\bar{\mathbf{A}}_b^T\mathbf{Q}_\varphi^{-1}\bar{\mathbf{A}}_b)^{-1} = \begin{pmatrix} \sigma_{B_h}^2 & \sigma_{B_h\dot{B}_h} & \sigma_{B_hB_v} & \sigma_{B_h\dot{B}_v} \\ \sigma_{\dot{B}_hB_h} & \sigma_{\dot{B}_h}^2 & \sigma_{\dot{B}_hB_v} & \sigma_{\dot{B}_h\dot{B}_v} \\ \sigma_{B_vB_h} & \sigma_{B_v\dot{B}_h} & \sigma_{B_v}^2 & \sigma_{B_v\dot{B}_v} \\ \sigma_{\dot{B}_vB_h} & \sigma_{\dot{B}_v\dot{B}_h} & \sigma_{\dot{B}_vB_v} & \sigma_{\dot{B}_v}^2 \end{pmatrix} \tag{4.12}$$

and can be visualised qualitatively by the error ellipses in figure 4.3. Their orientation can be obtained from the eigenvalues of $D\{\hat{\mathbf{b}}\}$. The constant component has its largest variance at an orientation $\beta = \bar{\theta}_0 \pm 90°$ with:

$$\tan\bar{\theta}_0 = \frac{\sigma_{\hat{B}_h\hat{B}_v}}{\sigma_{\hat{B}_h}^2 - \lambda_0} . \tag{4.13}$$

λ_0 is the major eigenvalue of $D\left\{\left(\hat{B}_h \ \hat{B}_v\right)^T\right\}$. The linear component has maximum variance for $\beta = \bar{\theta}_1$ and $\beta = \bar{\theta}_1 + 180°$, respectively, where

$$\tan\left(\bar{\theta}_1 + 90°\right) = \frac{\sigma_{\hat{\dot{B}}_h\hat{\dot{B}}_v}}{\sigma_{\hat{\dot{B}}_h}^2 - \lambda_1} . \tag{4.14}$$

λ_1 is the major eigenvalue of $D\left\{\left(\hat{\dot{B}}_h \ \hat{\dot{B}}_v\right)^T\right\}$, and $\bar{\theta}_1 \approx \bar{\theta}_0$ usually holds. The strong elongation of the error ellipses shows that the baseline is determined best perpendicular to the line of sight, whereas its rate of change has maximum precision in look direction of the sensor. This basically confirms the previously drawn conclusions identifying $\delta \dot{B}_\parallel$ and $\delta B_\perp$ as the components with the most significant impact on the interferometric phase.

Even though the complementary components $\delta B_\parallel$ and $\delta \dot{B}_\perp$ are theoretically estimable, their estimates would be too weakly determined to be considered reliable. This can be seen from figure 3.5a, where a relatively huge error in $B_\parallel$ induces only a very faint error signal in the phase. Conversely, a faint atmospheric signal that matches by chance this phase pattern, would result in unrealistically large estimates of $\delta B_\parallel$ in the order of metres (see figure 4.9a). Analogous considerations apply to $\delta \dot{B}_\perp$ (see figure 3.5d). Therefore, it is preferable to constrain these two components to zero. This can be achieved by narrowing the parameter space from four parameters $\mathbf{b}$ to two parameters $\mathbf{b}_\theta^T = (\, \delta \dot{B}_\parallel \;\; \delta B_\perp \,)$, yielding:

$$\mathrm{E}\{\boldsymbol{\varphi}\} = \bar{\mathbf{A}}_b \mathbf{T}^T \mathbf{b}_\theta \tag{4.15}$$

with:

$$\mathbf{T} = \begin{pmatrix} 0 & \sin(\theta_0) & 0 & -\cos(\theta_0) \\ \cos(\theta_0) & 0 & \sin(\theta_0) & 0 \end{pmatrix} . \tag{4.16}$$

The mean look angle θ_0, which is required for the decomposition into parallel and perpendicular component here, is heuristically defined by:

$$\theta_0 := \frac{\bar{\theta}_0 + \bar{\theta}_1}{2} . \tag{4.17}$$

The deviation between $\bar{\theta}_0$ and $\bar{\theta}_1$ depends on the spatial distribution of phase observations and is usually small, i. e., on the $0.1°$ level.

Least squares adjustment yields:

$$\hat{\mathbf{b}}_\theta = \left(\mathbf{T} \bar{\mathbf{A}}_b^T \mathbf{Q}_\varphi^{-1} \bar{\mathbf{A}}_b \mathbf{T}^T \right)^{-1} \mathbf{T} \bar{\mathbf{A}}_b^T \mathbf{Q}_\varphi^{-1} \boldsymbol{\varphi} \tag{4.18}$$

$$\mathrm{D}\{\hat{\mathbf{b}}_\theta\} = \sigma_0^2 \mathbf{Q}_\theta = \sigma_0^2 \left(\mathbf{T} \bar{\mathbf{A}}_b^T \mathbf{Q}_\varphi^{-1} \bar{\mathbf{A}}_b \mathbf{T}^T \right)^{-1} \tag{4.19}$$

with an estimable variance factor:

$$\hat{\sigma}_0^2 = \frac{\mathbf{v}_\varphi^T \mathbf{Q}_\varphi^{-1} \mathbf{v}_\varphi}{n_\varphi - u} , \tag{4.20}$$

where $u = 3$ is the number of unknowns ($\delta \dot{B}_\parallel$, $\delta B_\perp$ and φ_0). $\mathbf{v}_\varphi$ are the predicted corrections:

$$\mathbf{v}_\varphi = \bar{\mathbf{A}}_b \mathbf{T}^T \hat{\mathbf{b}}_\theta - \boldsymbol{\varphi} . \tag{4.21}$$

Note that the here addressed corrections denominate updates to the observations and are not identical with residuals. The notion of residuals rather refers to the remainder of the observations after subtraction of their predictions and thus implies an opposite sign.

4.3.2. Gridsearch Estimator

A major shortcoming of the least squares estimator outlined in the previous subsection is that it requires unwrapping. However, there are many applications in which unwrapping is cumbersome or even infeasible. In these cases, an alternative *gridsearch* approach can be pursued. It consists in minimising an objective function of the wrapped phase by incrementally searching the parameter space spanned by $\delta \dot{B}_\parallel$ and $\delta B_\perp$.

From eqs. (4.15) and (4.21) follows $\mathrm{E}\{\boldsymbol{\varphi} - \bar{\mathbf{A}}_b \mathbf{T}^T \mathbf{b}_\theta\} = 0$. By analogy to the ensemble coherence from (FERRETTI ET AL., 2001), a dedicated coherence measure can be defined as a function of $\mathbf{b}_\theta^T = (\, \delta \dot{B}_\parallel \;\; \delta B_\perp \,)$:

$$\gamma(\mathbf{b}_\theta) = \frac{1}{n_\varphi} \sum_{j=1}^{n_\varphi} e^{i(\varphi_j - \bar{\mathbf{a}}_{b,j} \mathbf{T}^T \mathbf{b}_\theta)} , \tag{4.22}$$

where i is the imaginary unit and $\bar{\mathbf{a}}_{b,j}$ is the jth row of $\bar{\mathbf{A}}_b$. Considering $0 \leq |\gamma| \leq 1$ and $\mathrm{E}\{|\gamma|\} = 1$, $\hat{\mathbf{b}}_\theta$ is defined as the set of parameters that maximises $|\gamma|$. Note that $\gamma(\mathbf{b}_\theta)$ can also be interpreted as a discrete integral transform of the two-dimensional signal $e^{i\varphi}$ to the $(\delta\dot{B}_\parallel, \delta B_\perp)$-domain.

As γ is insensitive to arbitrary cycle jumps of individual phases φ_j, application of the gridsearch estimator does not require explicit phase unwrapping. The computational load is higher than for the least squares method but still negligible compared to other InSAR processing steps. Whereas the *gridsearch* approach does not provide any intrinsic quality measures for the estimates, heuristic, peak-to-noise ratio-like indicators can be defined. A noteworthy drawback is that the estimates turn out to be unreliable in some cases, in particular when $|\gamma|(\mathbf{b}_\theta)$ has more than one distinct local maximum (see figure 6.5b).

4.4. Parameterisation

A central aspect of baseline error estimation is the choice of an optimal parametric model. This is always a trade-off between too few parameters causing approximation errors on the one hand and leakage of unmodelled signals due to overparameterisation on the other hand. It is thus considered appropriate to give increased attention to this conflict by evaluating the approximation quality for some common approaches and also discussing the drawbacks of overparameterisation.

For many conceivable parametric models, the parameter set can be subdivided into two subsets of parameters accounting for individual one-dimensional error signals in azimuth and range, respectively. For instance, the linear phase ramp approach ($p_1(\xi, \eta)$ in eq. (4.3)) involves a trend parameter a_{10} in azimuth and a trend parameter a_{01} in range. In the model with $(\delta\dot{B}_\parallel, \delta B_\perp)$, $\delta\dot{B}_\parallel$ represents a trend in azimuth, and $\delta B_\perp$ is equivalent to a trend in range. In both cases, the two parameters are stochastically decorrelated when estimated from homogeneously distributed observations. This justifies the separate evaluation of parametric models in azimuth and range, beginning with range.

4.4.1. Range Component

To evaluate parametric approximations of the orbital error signal $\delta\phi_{\mathrm{orb}}$, assumptions on the generating baseline error are required. An intrinsic difficulty of baseline error analysis is that the actual error is unknown and that general statistical characterisations suffer from the limitations elaborated in section 3.3.3. HANSSEN (2001, p. 121) assumes for his investigations an anisotropic error of $\delta B_v = 5\,\mathrm{cm}$ and $\delta B_h = 8\,\mathrm{cm}$. However, these quality measures from (VISSER ET AL., 1997; SCHARROO AND VISSER, 1998) refer to ERS orbit products from the 1990s. They are neither stationary in time and space nor do they result from a rigorously independent validation. Whereas the radial error component δB_v is indeed usually smaller than the across-track component δB_h, a general assumption of orbit accuracy for a whole mission and its adaption to InSAR applications is a delicate challenge.

To avoid arbitrary assumptions as far as possible, the following simulations are based on an error baseline $\delta\vec{B}$ (see figure 4.4) of constant length $\delta B = 10\,\mathrm{cm}$ and varying orientation β. The performance of an approximation model is evaluated by the *maximum deformation bias* $\Delta\delta D$, which is defined as maximum minus minimum absolute phase difference between error signal and approximation model (see figure 4.5), converted to its range equivalent:

$$\Delta\delta D = \frac{\lambda}{4\pi} \left(\delta\phi_{\mathrm{max}} - \delta\phi_{\mathrm{min}}\right) . \tag{4.23}$$

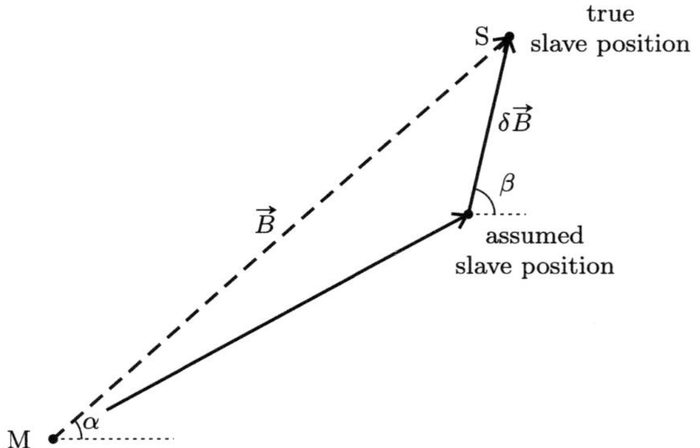

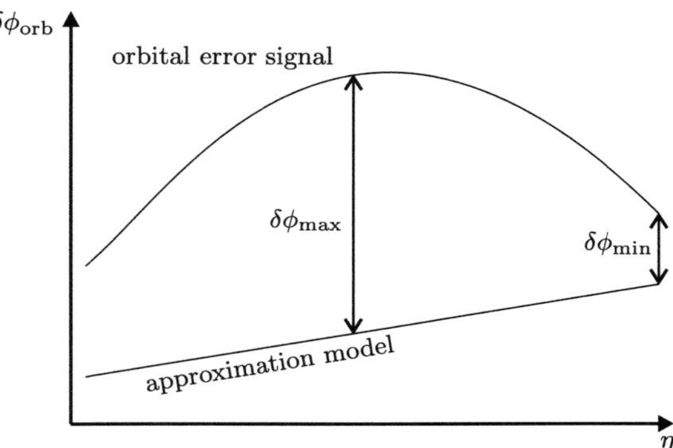

Figure 4.4.: The error baseline $\delta\vec{B}$ is defined as the difference vector between the true baseline and the assumed baseline as computed from orbit data. Besides the decomposition into the baseline error components $(\delta B_h, \delta B_v)$ and $(\delta B_\parallel, \delta B_\perp)$, respectively, it can also be fully characterised by its absolute value $\delta B = |\delta\vec{B}|$ and the orientation angle β. The visualisation in this figure assumes a deterministic master position M and a biased slave position S. (Figure based on HANSSEN, 2001, figure 4.12)

Figure 4.5.: The approximation error is defined as maximum absolute difference $\delta\phi_{\mathrm{max}}$ minus minimum absolute difference $\delta\phi_{\mathrm{min}}$ between the phases of an orbital error signal and its approximation by a parametric model. The phase error can be translated to a maximum deformation bias $\Delta\delta D$ of the measured LOS deformation according to eq. (4.23). Note that this definition is insensitive to a global phase offset, which has no influence on the interpretation of an interferogram.

This definition has the advantage that the error measure is insensitive to a constant phase offset. Whereas again the Envisat IS2 geometry is assumed for the sample computations within this section, numerical performance indicators for other sensors can be found in table A.4.

Considering initially the *uncorrected* orbital error signal, it can be seen from figure 4.6a that the maximum deformation bias depends on the orientation angle β of the error baseline. The maxima at $\beta = \theta$ and $\beta = \theta + 180°$ are in accordance with the conclusion from section 3.4.1 that the component $\delta B_\perp(\theta) = \delta B \cos(\theta - \beta)$ dominates the error signal in the phase. Accordingly estimating only one parameter $\delta B_\perp(\theta_0)$ from the simulated orbital error signal and correcting the interferogram by re-computing the reference phase reduces the maximum deformation bias by 98.6 % (from 10.9 mm to 0.16 mm; see figure 4.6). Of course, a full parameterisation of the two-dimensional baseline by $(\delta B_\parallel(\theta_0), \delta B_\perp(\theta_0))$ or (B_h, B_v), respectively, would completely eliminate the bias but also involves potential drawbacks of overparameterisation (see section 4.4.4).

Figure 4.6b also provides a comparison with the conventional approach of adjusting a polynomial (4.3) to the orbital phase, where $\xi = $ const for the one-dimensional evaluation of the range component. Approximation by a linear polynomial $p_1(\eta) = a_{00} + a_{01}\eta$ or phase ramp, respectively, yields a maximum bias of 0.59 mm, which means a reduction of 94.6 %. Hence, the one-parametric baseline approximation $\delta B_\perp$ performs for Envisat 3.8 times better than its one-parametric equivalent $p_1(\eta)$. (The auxiliary parameter a_{00} does not count.) On the other hand, a maximum deformation bias of 0.59 mm for the linear polynomial can be considered tolerable for a large number of applications. The maximum bias of a quadratic approximation $p_2(\eta) = a_{00} + a_{01}\eta + a_{02}\eta^2$ is only 0.07 mm (99.4 % improvement), which is smaller than the bias of the $\delta B_\perp$-parameterisation, but still outperformed by the likewise two-parametric approach of estimating the full two-dimensional baseline $(\delta B_\parallel, \delta B_\perp)$.

These conclusions may be different for other missions, for which a wider swath results in an increased nonlinearity of the orbital error signal. For Sentinel-1, the maximum bias of a linear approximation by $p_1(\eta)$ is as much as 2.87 mm (see table A.4) within the swath width of 250 km ground range. But also a $\delta B_\perp$-parameterisation performs worse (0.88 mm), because the sensitivity of the phase with respect

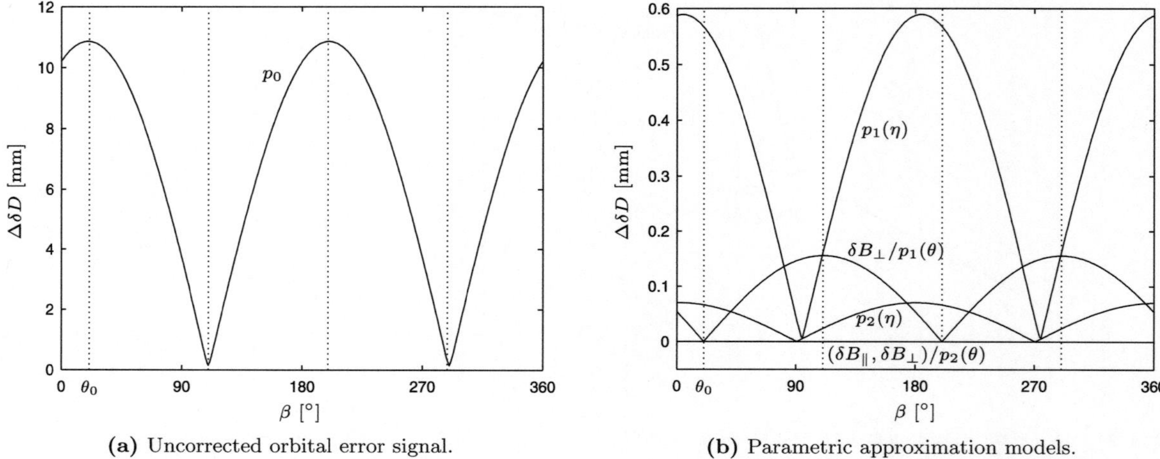

(a) Uncorrected orbital error signal. **(b)** Parametric approximation models.

Figure 4.6.: Maximum deformation bias of a simulated orbital error signal and performance of an approximation thereof. The simulations are based on an error baseline of $\delta B = 1\,\mathrm{dm}$ length with varying orientation β (see figure 4.4). Sensor height and field of view for a full Envisat IS2 interferogram are assumed (see table A.2). The plotted biases $\Delta\delta D$ are defined in eq. (4.23) and scale almost linearly with δB. Qualitatively, the results are approximately conferable to other sensors. For a quantitative evaluation see table A.4. **(a)** Maximum deformation bias of the uncorrected orbital error signal with respect to an unbiased interferogram with spatially constant phase. In terms of approximation, this error measure can also be interpreted as error of an approximation by a constant phase offset or a zero-degree polynomial, respectively; hence the symbol p_0. **(b)** Maximum deformation bias of the residual orbital error signal after approximation with dedicated parametric models. Mind the enlarged scale with respect to the graph on the left.

to the unmodelled $\delta B_\parallel$-component increases (i. e., the one-fringe equivalent $\delta B_{\parallel,2\pi}$ of an error in $B_\parallel$ significantly decreases; see table A.3).

All biases displayed in figure 4.6b are computed by rigorously simulating the phase error for a given orbit error and estimating the parameter or set of parameters, respectively, that minimises the sum of squared phase residuals. The bias is inferred from the range of residuals in accordance with eq. (4.23). The approach pursued in (BÄHR AND HANSSEN, 2012, figure 3b) was identical except for the $\delta B_\perp$-model, where instead the bias due to ignoring the $\delta B_\parallel$ component has been evaluated by comparing the phase signals of the assumed baseline error $\delta\vec{B}$ and a projected baseline error $\langle\delta\vec{B},\vec{r}_\mathrm{M}\rangle\vec{r}_\mathrm{M}$. This comparison is considered less meaningful, because the true baseline error and its projected components are generally unknown in practical applications.

BÄHR AND HANSSEN (2012) further proposed the alternative to approximate the two-dimensional baseline error by a polynomial in look angle:

$$p_d(\theta) = \sum_{k=0}^{d} a_k \left(\frac{\theta - \theta_0}{\Delta\theta}\right)^k . \tag{4.24}$$

Whereas the models:

$$p_d(\eta) = \sum_{k=0}^{d} a_k \left(\frac{\eta - \eta_0}{\Delta\eta}\right)^k \quad \text{and} \quad p_d(R) = \sum_{k=0}^{d} a_k \left(\frac{R - R_0}{\Delta R}\right)^k \tag{4.25}$$

are equivalent since $\partial R/\partial\eta = 1$, the approximation by $p_d(\theta)$ is indeed different, because $\partial\theta/\partial\eta$ is not constant. A linear polynomial in θ performs significantly better than a linear polynomial in η (see figure 4.6b), but the approach does not yield an improvement with respect to the $\delta B_\perp$-parameterisation. Compared rigorously, the maximum deformation biases of $p_1(\theta)$ and $\delta B_\perp$ show only very small mutual deviations, and the bias of $p_2(\theta)$ is practically zero. Nevertheless, even if there is no benefit in accuracy of the polynomial in θ, an individual computation of $\theta(\eta)$ for every pixel may be easier to implement than the recomputation of the reference phase with an updated baseline.

To support the choice of an adequate model, table A.4 lists the maximum deformation biases for different sensors and modes, consistently assuming an error baseline of constant length $\delta B = 1\,\mathrm{dm}$. All numbers can be adapted to bigger or smaller baseline errors, since the phase error scales linearly with dB. This is evident from eq. (3.15) under consideration of eq. (3.7).

For the orbital error signal in range it can be concluded that in case of unexceptional orbit errors below $1\,\mathrm{dm}$ a $p_1(\eta)$-correction is sufficient. For larger errors, a direct parameterisation of the baseline error component $\delta B_\perp$ should be considered. This applies especially to Radarsat-2 and even more to Radarsat-1, for which the orbit data are generally more inaccurate. Also for the planned mission Sentinel-1 with a swath width of 250 km in the IWS mode, it is advisable to rely on the $\delta B_\perp$-parametrisation.

4.4.2. Azimuth Component

The error signal in azimuth is a direct translation of variations in $\delta B_\parallel$ into the interferometric phase (see figure 4.2). For instance, a constant rate of change $\delta\dot{B}_\parallel$ induces a perfectly linear error signal. Generally, any phase pattern $\delta\phi_{\mathrm{orb}}(t)$ could result from an arbitrarily varying baseline error $\delta B_\parallel(t)$. However, as orbital trajectories have the nature to be very smooth curves, it is unlikely that errors in these curves undergo completely random variations. Therefore, a linearly varying error in the parallel baseline component is considered an appropriate parametrisation of the baseline error for the short acquisition time of a single radar scene (Envisat: 15 s). Higher order polynomials involve an increased risk of overparameterisation (see section 4.4.4). Eventually, a quadratic polynomial may be considerable for long swath processing.

Due to the approximately linear relation (4.2) between errors in the parallel baseline and their phase contributions, an approximation polynomial in ξ is almost equivalent with a polynomial of equal degree in $\delta B_\parallel$. In numerical simulations within this thesis, where a spherical reference surface below a concentric spherical orbit is assumed, no difference can be observed. In practical applications, negligible minor discrepancies may occur though. PEPE ET AL. (2011, figure 2d) for instance, observed a small residual phase signal of higher order after approximating an error signal due to a linear variation in $B_\parallel$ by a linear polynomial in ξ.

4.4.3. Influence of Topography

So far, parametric models have only been evaluated for the spherical reference surface of a flat earth. However, it has been shown in figure 3.7 that the orbital error signal can expose distinct nonlinear features in the presence of significant topographic variations. As a consequence, the performance of parametric models may decrease with increasing height variations. This effect is analysed in the following.

As in section 4.4.1, an error baseline of constant length δB is simulated for various orientations β and kept constant over azimuth. The corresponding error signal is simulated for different generic topographic patterns (see figure 4.7a) spanning a full Envisat IS2 scene, and its approximation error is evaluated for three parametric models: baseline error components $(\delta\dot{B}_\parallel, \delta B_\perp)$, a linear phase ramp $p_1(\xi, \eta)$ and a quadratic polynomial $p_2(\xi, \eta)$. The performance of the approximation is characterised by the maximum deformation bias (4.23), subsequently maximised with respect to the error baseline orientation β (see figure 4.7b-c).

The model relying on the baseline error parameters $(\delta\dot{B}_\parallel, \delta B_\perp)$ does not perform significantly worse in presence of topography. This is evident, because the DEM information is fully considered for the

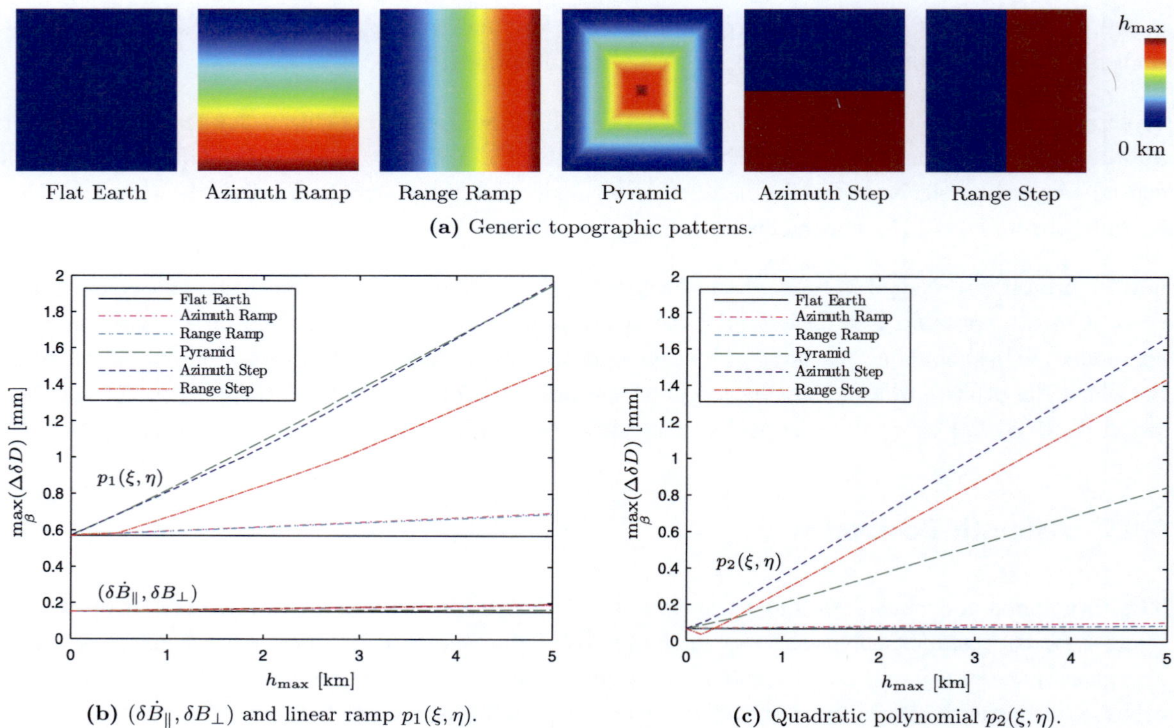

(a) Generic topographic patterns.

(b) $(\delta\dot{B}_{\parallel}, \delta B_{\perp})$ and linear ramp $p_1(\xi, \eta)$.

(c) Quadratic polynomial $p_2(\xi, \eta)$.

Figure 4.7.: Performance of approximating the orbital error signal by different parametric models in the presence of significant topographic variations. Simulations are based on an error baseline of $\delta B = 1\,\mathrm{dm}$ length with varying orientation β that is constant over azimuth. Sensor height and field of view for a full Envisat IS2 interferogram (see table A.2) are assumed. Qualitatively, the results are approximately conferable to other sensors. For a quantitative evaluation, please refer to table A.4.

computation of a corrected reference phase from a corrected baseline. The polynomial approximations on the other hand do not exploit any topographic information when fitting a ramp, paraboloid or surface of higher order to the interferogram. Consequently, the approximation quality degrades if the range of topographic heights increases, the deformation bias increasing more or less proportionally with the height scale $h_{\max}$. As the approximation error also depends on the individual topography of the area of interest, it is not feasible to give a general estimate of the maximum possible deformation bias. Nonetheless, the evaluations with different generic DEMs in figure 4.7 provide a rough idea of expectable biases in Envisat interferometry. An assessment for other missions can be supported by table A.4, where the maximum deformation bias is computed for a pyramidal DEM with $h_{\max} = 1\,\mathrm{km}$.

Baseline variations in azimuth are ignored by these evaluations for several reasons. First of all, there is no natural functional representation of the temporal variations of a two-dimensional error baseline that could be assumed for the simulations. Even the most self-evident choice, a linearly varying error baseline, would not alter the observed approximation error significantly, because errors in $\dot{B}_{\perp}$ have only a small effect on the phase (see figure 3.5d), and errors in $\dot{B}_{\parallel}$ would be fully absorbed by the linear parameters $\delta\dot{B}_{\parallel}$ or a_{10}, respectively, of the approximation models. Analogous conclusions can be drawn for higher order coefficients.

Whereas the parameterisation of the orbital error signal by a linear ramp $p_1(\xi, \eta)$ is still a sufficiently good approximation for many applications, it can involve very significant biases if large topographic variations go along with large baseline errors. For instance, PEPE ET AL. (2011, figure 7) observe deformation biases on the decimetre level for a ramp-corrected Radarsat-1 interferogram of the island of Hawaii.

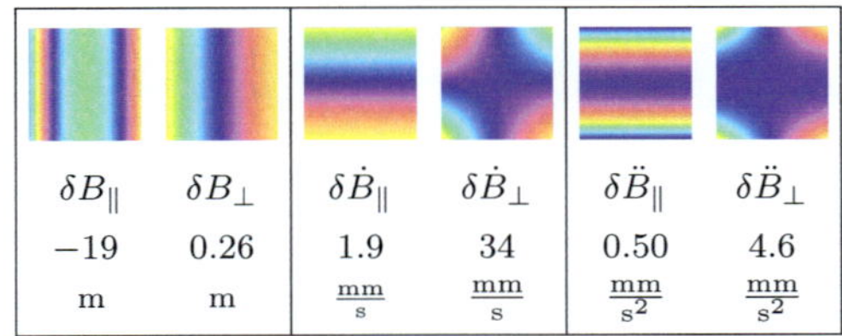

(a) Baseline error parameterisation.

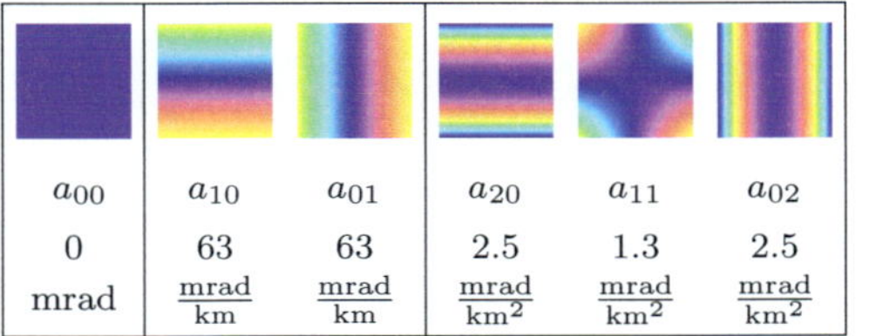

(b) Polynomial representation (4.3).

Figure 4.8.: Signal spaces of individual parameters commonly used to approximate the orbital error signal, simulated for the geometry of an Envisat IS2 interferogram.

Nevertheless, the applicability of the phase ramp approach also depends on error budget and accuracy requirements of the respective application.

4.4.4. Overparameterisation

Mutual inference of baseline errors $\delta \vec{B}(t)$ and associated orbital error signals $\delta\phi_{\mathrm{orb}}(\xi, \eta)$ involves some important restrictions. On the one hand, not every phase pattern can be explained by baseline errors. All physically possible error signals lie within an accordant signal space, subspaces of which can be associated with individual baseline components (see figure 4.8a). On the other hand, baseline error estimation requires an appropriate choice of parameters, because it is impossible to fully describe the continuously varying baseline error by a finite parameter set. A too small number of parameters results in approximation errors, which have been discussed in the previous subsections. Increasing the number of parameters on the other hand runs the risk of incorporating poorly significant parameters.

A poorly significant parameter (e. g., $\delta B_{\parallel}$ or $\delta \dot{B}_{\perp}$) qualifies by a low phase sensitivity, meaning that a very large baseline error is required to induce a perceivable error pattern in the phase. Conversely, the estimation of this parameter from the phase is an ill-posed problem, implying that a very faint, non-orbital signal in the phase that matches by chance the parameter's signal space, can yield an extraordinary large baseline error. These mechanisms have two major drawbacks (see figure 4.9):

1. Unmodelled atmospheric or deformation signals can generate irrealistically large baseline error estimates. However, if the objective of baseline error estimation is mitigating its effect on the phase rather than quantifying the error itself, physically irrealistic estimates can be tolerated.

2. Unmodelled atmospheric or deformation signal components may be attributed to the orbital contribution. If deformation is the signal of interest, this is uncritical for the atmospheric but not for the deformation component. The leakage can result in a distorted deformation signal and consequentially in biased deformation estimates.

As biased deformation estimates should be avoided as far as possible, it suggests itself to limit the parameter set to a few significant and well-estimable baseline error components. This confines leakage of deformation signals into the orbital component, whereas the neglect of poorly significant parameters is a tolerable loss, given the little effect these parameters have on the phase. Analogous conclusions can be drawn for the polynomial approximation (4.3) and its signal space in figure 4.8b, where a limitation of the parameter set likewise confines leakage of deformation signals.

In conclusion, the choice of a parameter set is always a trade-off between minimising approximation errors and confining leakage of deformation signals into the orbital contribution. Two parameters like

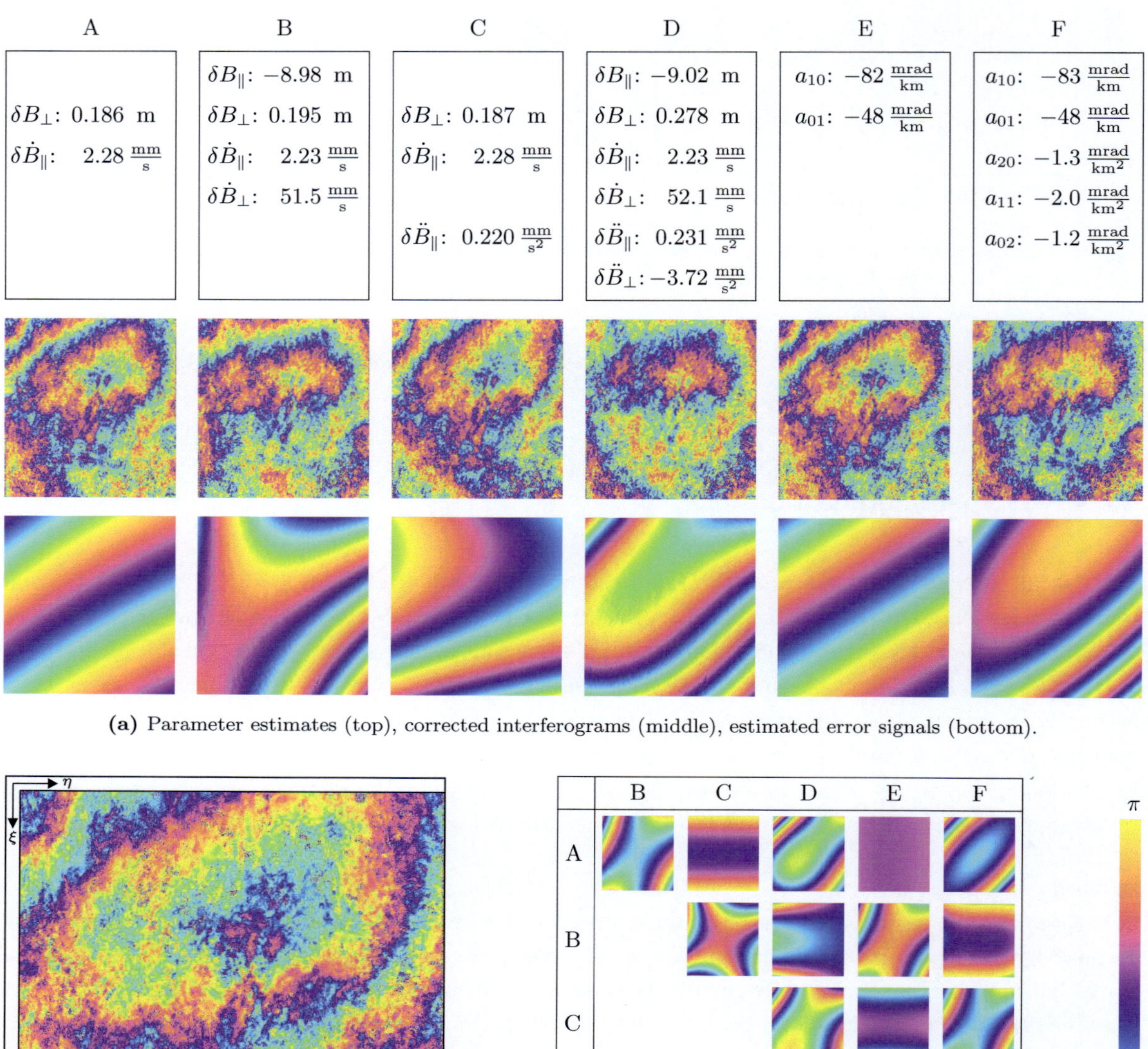

(a) Parameter estimates (top), corrected interferograms (middle), estimated error signals (bottom).

(b) Uncorrected interferogram.

(c) Differences of estimated error signals.

Figure 4.9.: Evaluation and comparison of different parameterisations for the estimation of baseline errors from an Envisat IS2 interferogram with a strong atmospheric contribution (interferogram 12281–14285 from the data set analysed in chapter 6). **A:** $(\delta\dot{B}_\parallel, \delta B_\perp)$ is a stable parameterisation; both $\delta B_\perp = 0.186$ m and $\delta\dot{B}_\parallel = 2.28$ mm/s are relatively large but still reasonable errors. **B:** A baseline error component of $\delta B_\parallel = -8.98$ m is very unlikely for the Envisat orbit. So is $\delta\dot{B}_\perp = 51.5$ mm/s, which would imply an error increase of 78 cm during the acquisition of a single scene (15 s). **C:** $\delta\ddot{B}_\parallel = 0.220$ mm/s^2 can be interpreted as an error variation from 6.2 mm over -6.2 mm back to 6.2 mm during the 15 s of acquisition, which is in a reasonable order of magnitude. However, the plausibility of quadratic components is difficult to assess, because the baseline error accumulates rapidly when extrapolated beyond the extent of one scene. **D:** $\delta\ddot{B}_\perp = -3.72$ mm/s^2 suggests an error variation from -10.6 cm over 10.6 cm back to -10.6 cm during 15 s, which becomes implausible when this local trend is smoothly extrapolated. **E:** The "linear phase ramp" is almost identical with the estimated error signal A and can thus be considered a stable parameterisation. **F:** This approximation is very similar to the signal D, which involves implausible estimates. Consequently, the applicability of quadratic polynomials is questionable.

$(\delta \dot{B}_\parallel, \delta B_\perp)$ or (a_{10}, a_{01}), respectively, are a minimum requirement for an appropriate parameterisation. The associated leakage of linear deformation signals is unavoidable. The benefit of an extension by a third parameter $\delta \ddot{B}_\parallel$ is at least arguable and eventually worth considering for processing long swath data that covers more than one frame in azimuth.

4.5. Reliability of Estimates

Selecting the phases of a particular subset of pixels as observations for baseline error estimation does not necessarily result in a homogeneous spatial distribution. As an unbalanced distribution can have unintended effects on the estimates, the estimation algorithm needs to be as insensitive as possible to the geometry of observations. But also resistance with respect to outliers is a critical aspect that is worth some discussion.

4.5.1. Spatial Distribution and Weighting of Observations

Theoretically, all available pixels could be considered in eq. (4.15) for the estimation of baseline error parameters. However, in most interferograms not every pixel can supply a reliable phase information. In regions of poor coherence, the interferometric phase is practically meaningless and cannot be exploited. However, it is not advisable to simply exclude poorly coherent pixels below a fixed coherence threshold, since this can lead to inhomogeneous spatial distributions of observations. Thus, local nuisance signals due to unmodelled contributions of atmospheric effects or deformation can exert a dominating influence on the estimates.

The most intuitive approach to avoid such "leverage" effects is to enforce a homogeneous spatial distribution of observations by defining a grid on the interferogram and picking from each grid cell only the pixel with the largest coherence (see figure 4.10). Nevertheless, this approach reaches its limits if not all grid cells contain pixels exceeding a minimum coherence threshold, below which a reliable phase measurement is very unlikely. But by disregarding grid cells with lack of reliable measurements, the homogeneity of the distribution is disrupted by more or less sporadic gaps.

A conceivable remedy for spatial imbalances of observations would be to compensate the local lack of coherent pixels by increasing the weights of surrounding pixels. Whereas this approach would ensure a spatially homogeneous distribution of weights, it could also involve exceptionally high weights for isolated observations in regions of poor coherence, associated with a strong influence on the unknowns and reduced potential for mutual validation with neighbouring observations. These deficiencies could eventually be overcome by a balanced adjustment (JURISCH AND KAMPMANN, 1998), applying an adaptive weighting scheme that equalises the influence of all observations on the estimates. A rigorous evaluation of this approach would require a thorough investigation though.

Another conceivable motivation for weighting observations is the dependence of the phase standard deviation on coherence as an indicator for decorrelation noise (HANSSEN, 2001, pp. 93-96). However, a coherence-based weighting may result in an inhomogeneous spatial distribution of weights, which is comparable to inhomogeneously distributed observations and can likewise trigger leverage effects. Hence, a weighting of observations should better be avoided.

In view of all aforementioned trade-offs and limitations, the following strategy is proposed for generation, selection and weighting of observations (see figure 4.10a-c,g): In the context of small baseline processing,

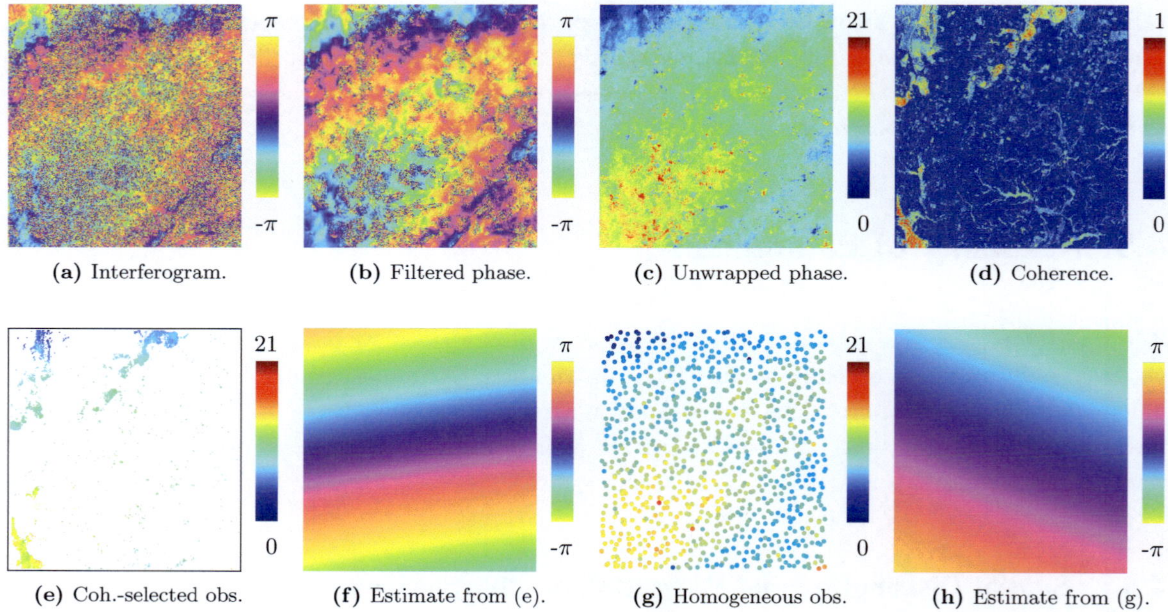

(a) Interferogram. (b) Filtered phase. (c) Unwrapped phase. (d) Coherence.

(e) Coh.-selected obs. (f) Estimate from (e). (g) Homogeneous obs. (h) Estimate from (g).

Figure 4.10.: Demonstration of leverage effects for a sample interferogram with a dominant atmospheric signal. It is shown that different observation selection strategies – (e) versus (g) – can produce significantly different estimates. **(a)** Interferogram 24305–28814 from the data set analysed in chapter 6 ($B_\perp = -378\,\mathrm{m}$, $B_T = 315\,\mathrm{d}$). Multilooking factors $m_\xi = 25$ and $m_\eta = 5$ yield 1072 multilooked pixels in azimuth and 1032 multilooked pixels in range, respectively. **(b)** Strong phase filtering (see section 2.3.5) facilitates phase unwrapping. **(c)** After phase unwrapping, the bias due to filtering is removed. **(d)** Sporadic salt lakes with persistent reflectivity characteristics cause an unbalanced distribution of coherent areas. **(e)** 20696 pixels with a coherence above 0.6. Neighbouring pixels have been aggregated for better visualisation. **(f)** Orbital error signal estimated from the observations displayed in image (e): $\delta\hat{B}_\parallel = -2.03\,\mathrm{m/s}$ and $\delta\hat{B}_\perp = -4.0\,\mathrm{cm}$. Locally clustered observations lever out the estimation, and the local phase trend from the upper-left corner dominates the estimated signal. **(g)** 1023 homogeneously distributed pixels serving as observations. They have been selected by picking the most coherent pixel from tiles of 30×30 multilooked pixels, disregarding 33 pixels with a coherence below 0.2. Pixels are represented by enlarged dots for better visualisation. **(h)** Orbital error signal estimated from the observations displayed in image (g): $\delta\hat{B}_\parallel = -1.08\,\mathrm{mm/s}$ and $\delta\hat{B}_\perp = 5.3\,\mathrm{cm}$. Leverage effects of local trends have been avoided by enforcing a homogeneous distribution of observations. After rejection of six outliers by data snooping, the estimates become $\delta\hat{B}_\parallel = -1.07\,\mathrm{mm/s}$ and $\delta\hat{B}_\perp = 5.5\,\mathrm{cm}$.

the interferogram is multilooked to reduce decorrelation noise. Strong phase filtering (see section 2.3.5) can optionally be applied to bridge small decorrelated patches and thus help preventing phase jumps during unwrapping. To avoid a bias due to filtering, the original phase is restored after unwrapping, albeit this bias is probably negligible for baseline error estimation. Observations are selected by picking the most coherent pixel from each cell of a regular grid defined on the interferogram. Finally, the pixels that do not fulfil a minimum coherence requirement are discarded at the expense of admitting sporadic gaps. In the context of PS processing, the selection procedure is the same, but neither multilooking nor phase filtering is applied, and only PS candidates qualify as observations.

4.5.2. Robustness and Data Snooping

An intrinsic requirement of reliable estimates is their resistance to outliers in the observations. In the context of baseline error estimation, two types of outliers can be distinguished: individual pixels not matching their surroundings and connected patches of pixels that are wrongly unwrapped as a whole. Whereas the first type can easily be detected by dedicated test statistics, the second type is cumbersome to identify, because erroneously unwrapped pixels mask each other if they are locally clustered. Considering further that phase observations are always relative, it is even impossible to determine for two equally sized halves of an interferogram, which one of the two is unwrapped correctly and which one is not. In light of these challenges, the detection of patch-wise unwrapping errors will only be considered for an interferogram as a whole by exploiting multiple interferometric combinations (see section 5.3).

Possible causes for outliers in individual pixels are unwrapping errors, locally confined deformation and pixel-wise, processing-induced artefacts. Especially the latter can produce outliers that lie very far off the global trend and entail seriously biased estimates. For their detection and elimination within the framework of the least squares estimator from section 4.3.1, BAARDA's (1968) *data snooping* is proposed. The underlying idea of this procedure is to statistically test observations individually on agreement with the mathematical model and to iteratively reject outliers until all tests pass. Whereas Baarda's approach requires that the stochastic model of the observations is entirely known, this is not the case in eq. (4.9), where an a priori unknown variance factor σ_0^2 is allowed for. But the idea of data snooping can also be conferred to dedicated data-adaptive tests like the τ-test (POPE, 1976) or the equivalent t-test (HECK, 1981; JÄGER ET AL., 2006, p. 193), that follow the τ- or the more common t-distribution, respectively.

Assuming normally distributed observations, the test statistic of the t-test for the ith phase observation φ_i follows a t-distribution (JÄGER ET AL., 2006, p. 188):

$$T_{\varphi,i} = \frac{\mathbf{v}_\varphi^T \mathbf{Q}_\varphi^{-1} \mathbf{e}_i \left(\mathbf{e}_i^T \mathbf{Q}_\varphi^{-1} \mathbf{Q}_{\mathbf{v}_\varphi} \mathbf{Q}_\varphi^{-1} \mathbf{e}_i \right)^{-1} \mathbf{e}_i^T \mathbf{Q}_\varphi^{-1} \mathbf{v}_\varphi}{\bar{\sigma}_0} \sim t_{n_\varphi - u - 1} \, , \tag{4.26}$$

where $\mathbf{e}_i$ is a column vector of zeros with a one at the ith position. $\mathbf{Q}_{\mathbf{v}_\varphi}$ is the cofactor matrix of the corrections $\mathbf{v}_\varphi = (v_{\varphi,i})$:

$$\mathbf{Q}_{\mathbf{v}_\varphi} = \mathbf{Q}_\varphi - \bar{\mathbf{A}}_b \mathbf{T}^T \mathbf{Q}_\theta \mathbf{T} \bar{\mathbf{A}}_b^T \, . \tag{4.27}$$

Assuming that there is indeed a blunder in the ith observation, the variance factor estimate $\hat{\sigma}_0^2$ from eq. (4.20) would be biased. Thus, the contribution of the ith correction is removed from the sum of squared corrections, yielding the adapted estimate:

$$\bar{\sigma}_0^2 = \frac{\mathbf{v}_\varphi^T \mathbf{Q}_\varphi^{-1} \mathbf{v}_\varphi - \mathbf{v}_\varphi^T \mathbf{Q}_\varphi^{-1} \mathbf{e}_i \left(\mathbf{e}_i^T \mathbf{Q}_\varphi^{-1} \mathbf{Q}_{\mathbf{v}_\varphi} \mathbf{Q}_\varphi^{-1} \mathbf{e}_i \right)^{-1} \mathbf{e}_i^T \mathbf{Q}_\varphi^{-1} \mathbf{v}_\varphi}{n_\varphi - u - 1} \, . \tag{4.28}$$

If $T_{\varphi,i}$ exceeds both a dedicated threshold deduced from the t-distribution and any other $T_{\varphi,j}$ with $j \in \{1, \ldots, n_\varphi\} \backslash i$, the ith observation is rejected, and the parameters are re-estimated from the remaining observations. This procedure is repeated until all test statistics fall below the threshold. It has to be taken care that the iterative rejection does not result in an unbalanced spatial distribution of observations, which could entail leverage effects as addressed in section 4.5.1. Consequently, the number of rejectable observations should be limited.

Beyond the potential of iterative rejection, iterative reweighting of observations within the framework of robust adjustment may have some potential to provide a more optimal outlier handling. By iteratively downweighting observations with excessive corrections, the influence of large groups of outliers can be bounded. A successful application of this concept would require further and thorough research to yield a generally applicable procedure that ensures reliable results. For further reading can be recommended: introductory reviews in the context of geodetic applications (KOCH, 1996, 1999; NIEMEIER, 2008), textbooks from mathematical statistics (HUBER, 1981; HAMPEL ET AL., 1986; ROUSSEEUW AND LEROY, 1987) as well as the dissertation of WICKI (1998). A promising application of robust adjustment has already been presented by SHIRZAEI AND WALTER (2011), who mitigate leakage of deformation signals by spatially low-pass filtering the interferogram prior to robust estimation of the orbital error signal. Since the filtering is implemented by a wavelet decomposition, their approach is not compatible to sparsely distributed observations though.

4.6. Stochastic Model

Unbiased baseline error estimates and reliable quality measures thereof require a consistent estimator, for which it is assured that all contributions to the residual interferometric phase φ are either negligible or appropriately considered in the mathematical model. For the least squares estimator, the latter subsumes the functional model (4.15) and the stochastic model (4.9). Whereas the functional model is well-defined by the objective of baseline error estimation, the stochastic model has to accommodate contributions from each of the five signal components distinguished in figure 2.7:

- $\delta\phi_{\mathrm{topo}}$: Topographic artefacts, i. e., errors in the assumed reference surface (see figure 2.2) can have different causes. They may be due to the general DEM uncertainty, systematic inconsistencies between the radar penetration depths and the DEM reference surface or interpolation errors resulting from incongruent sampling of DEM and SAR images. The induced error signals are either uncorrelated or have relatively short correlation lengths. Global offsets or tilts of the whole DEM are of minor concern, because these kinds of error are easily detectable and correctable outside the framework of InSAR processing.

- $\delta\phi_{\mathrm{orb}}$: Orbit errors are almost completely absorbed by the dedicated baseline error parameters. Remaining orbital signal components that do not match the corresponding signal spaces (see figure 4.8a) have relatively long spatial wavelengths.

- ϕ_{defo}: Deformation is assumed to be either non-existent (e. g., if the observations φ_i have been selected from non-deforming regions) or estimated in a separate algorithm, iteratively alternating with baseline error estimation. In any case, residual deformation signals may remain, which are of small amplitude and spatially correlated.

- ϕ_{atmo}: For the atmospheric contribution, different components are distinguished (see section 2.2.3). The turbulent component can be considered a stationary, spatially correlated stochastic process with a relatively short correlation length. Large-scale gradients of temperature or pressure are partly absorbed by the orbit error parameters in the functional model. Stratigraphic effects are assumed to be either negligible or accounted for otherwise, for instance by extending the functional model by a height regression parameter.

- ϕ_{noise}: Measurement noise, processing noise and effects of decorrelation behave randomly and are spatially uncorrelated. Coherence estimates (2.11) of individual pixels enable a discrimination between more or less precise measurements.

It follows that a rigorous stochastic model would have to accommodate three classes of signals: uncorrelated noise (residual topographic errors, measurement noise, processing noise, decorrelation), signals with short and medium correlation length (topographic errors, residual deformation, turbulent atmosphere) and large-scale signals (residual orbital signals, residual deformation, atmospheric gradients). However, the establishment of an all-comprehensive stochastic model is not feasible without a reasonable effort, and compromises have to be made.

The uncorrelated noise can straightforwardly be modelled by a diagonal covariance matrix. But associating more or less decorrelated pixels with individual variances as a function of coherence would be in conflict with the requirement of homogeneous weights to avoid leverage effects (see section 4.5.1). Thus, all variances are homogenised, and the most simplistic model to be evaluated is a scaled identity matrix:

$$\mathrm{D}\{\boldsymbol{\varphi}\} = \sigma_0^2 \mathbf{I} \,. \tag{4.29}$$

Among signals with short or medium correlation length, the contribution of the turbulent atmosphere clearly predominates topographic errors and residual deformation that may both be insignificant. The stochastic properties of atmospheric turbulence can be adequately described by a dedicated covariance function $C(\Delta\xi, \Delta\eta)$ (KNOSPE AND JÓNSSON, 2010), which is often simplified by an isotropic model $C(r)$ (HANSSEN, 2001, p. 144), where $r \propto \sqrt{\Delta\xi^2 + \Delta\eta^2}$. As the characteristics of the atmospheric signal can be very different (HANSSEN, 2001, pp. 140 et seq.), covariance functions have to be adjusted to the individual interferograms. Thus, a more accurate stochastic model consists of a fully populated covariance matrix:

$$
\mathrm{D}\{\boldsymbol{\varphi}\} = \sigma_0^2 \begin{pmatrix} C(0) & C(r_{12}) & \cdots & C(r_{1n_\varphi}) \\ C(r_{12}) & C(0) & \cdots & C(r_{2n_\varphi}) \\ \vdots & \vdots & \ddots & \vdots \\ C(r_{1n_\varphi}) & C(r_{2n_\varphi}) & \cdots & C(0) \end{pmatrix}, \tag{4.30}
$$

where r_{ij} is the spatial distance between pixels i and j. The scaling factor σ_0^2 concedes a degree of freedom to residual inconsistencies.

Finally, large-scale signals that are not covered by the signal space of the baseline error model constitute a serious problem, because an adequate stochastic modelling is not feasible. If they were attributed to a stochastic process with correlation lengths larger than the interferogram, the characteristics of this process could not be captured from the available data. So far, the most practicable solution is to neglect these signals in modelling.

In conclusion, the stochastic model of the least squares approach involves several compromises. The major deficiency is probably the neglect of large-scale non-orbital contributions; but also the homogenisation of observation variances, atmospheric anisotropy as well as non-stationarity of topographic errors and residual stratigraphic effects contribute to an inconsistent methodology. The inconsistency of the gridsearch estimator is disproportionately larger as it does not provide a stochastic model at all.

4.7. Conclusions

Chapter 4 has provided a compilation of approaches for estimating relative orbit errors from the phases of a single interferogram. A comparative assessment has primarily focussed on aspects such as parameterisation and observation design. It turned out that the approximation error of a first order polynomial in azimuth and range ("linear phase ramp") is insignificant in case of unexceptional orbit errors below one decimetre and flat to moderate topography. Higher order parameterisations run a serious risk of overparameterisation except for long swath processing. The mathematical framework has been provided for two estimators using an optimal $(\delta\dot{B}_\parallel, \delta B_\perp)$-parameterisation. It has been shown that the observation design, i. e., the spatial distribution and weighting of phase observations, can be critical for obtaining unbiased estimates and still has potential for optimisation. As this would be a research topic on itself, a preferably homogeneous distribution of unweighted phase measurements has been proposed as an appropriate compromise to minimise related biases. Finally, the choice of a stochastic model has been addressed. Acknowledging that rigorous modelling is far from straightforward, it has been considered to neglect all contributions except the usually dominating atmospheric turbulence, which can be described by a distance-dependent covariance function.

An actual evaluation of the proposed estimators is most valuable when additionally exploiting the potential of cross-validation with different linear combinations of images. A dedicated framework of network adjustment is presented in chapter 5 and evaluated in chapter 6, where it is applied to a real data set.

5. Network Adjustment of Orbit Errors

Residual fringe patterns in an interferogram allow to infer only relative but not absolute orbit corrections. Applying these corrections means adjusting the trajectories of master and slave acquisition with reference to each other. Hence, orbit error estimation from one interferogram alone is equivalent to the estimation of baseline errors, which has been subject to the previous chapter.

As interferograms with suboptimal coherence may be contaminated by unwrapping errors, it is essential to ensure the reliability of baseline error estimation. A promising approach is to cross-check individual baseline error estimates by exploiting linear combinations of interferograms with different perpendicular and temporal baselines. Generalising this concept, a network of interferograms is set up, connecting the available images on redundant paths in the spatio-temporal baseline-space $(B_\perp, B_T)$. By adjusting the individual baseline error estimates of all interferograms in the network, not only a mutual validation is provided, but also quasi-absolute, image-wise orbit errors can be inferred. These either refer to a global master image or satisfy a minimum-norm condition. Network adjustment also involves an enhancement of precision by adjusting inconsistencies due to interferogram-specific filtering. Furthermore, orbit error estimates for poorly coherent interferograms can be improved or even enabled by inference from multiple adjacent interferometric combinations.

This chapter provides the mathematical framework for a network adjustment of orbit errors parameterised by the baseline error components $\delta\dot{B}_\parallel$ and $\delta B_\perp$. Its evaluation and performance assessment is postponed to chapter 6, where conclusions are drawn from application to a real data set.

5.1. Classification of Approaches

A couple of different approaches have been proposed so far for network adjustment of orbit errors (KOHLHASE ET AL., 2003; BIGGS ET AL., 2007; PEPE ET AL., 2011; BÄHR AND HANSSEN, 2012). In addition to the aspects discussed in section 4.2, distinctions can be made in the following categories:

1. *Parameter-wise partitioning.* The parameters corresponding to signals in either azimuth or range can be estimated separately (KOHLHASE ET AL., 2003; PEPE ET AL., 2011) or jointly (BIGGS ET AL., 2007; BÄHR AND HANSSEN, 2012). In case of homogeneously distributed phase observations in the regular spherical model geometry (see appendix A), the two sets are uncorrelated. In reality, there may be some correlations due to an inhomogeneous distribution of observations, but these are mostly small enough to be negligible. Nevertheless, seeing the manageable computational complexity of a simultaneous adjustment, no significant benefit can be drawn from partitioning.

2. *Hierarchical organisation.* Orbit error parameters can be estimated either in a closed one-step adjustment or in a sequential two-step approach. Closed adjustment means estimating acquisition-specific orbit errors directly from the individual phase measurements (KOHLHASE ET AL., 2003; BIGGS ET AL., 2007). The sequential approach involves intermediate interferogram-specific baseline error estimates to which acquisition-specific orbit errors are adjusted in a second step (PEPE ET AL.,

2011; BÄHR AND HANSSEN, 2012). The two concepts are specified in detail in section 5.2.1 and comparatively evaluated in section 6.2.4.

3. *Datum definition.* The most generic approach to define an absolute reference for the estimated errors is the introduction of a minimum-norm condition (KOHLHASE ET AL., 2003; BIGGS ET AL., 2007; BÄHR AND HANSSEN, 2012). PEPE ET AL. (2011) also acknowledge this option but estimate their orbit errors relative to an assumedly deterministic master orbit. Also intermediate datum definitions involving partial trace minimisation can be practicable (BÄHR AND HANSSEN, 2012). A detailed discussion is provided in section 5.2.6.

It is interesting to note that network approaches are also applicable in the context of the atmospheric signal. ELLIOT ET AL. (2008) extend the approach of BIGGS ET AL. (2007) by a network atmospheric correction, estimating one parameter per acquisition that accounts for the correlation of the topographic height and the stratigraphic atmospheric component. LIU ET AL. (2010) perform a pixel-wise network adjustment of the atmospheric contribution, assuming zero deformation and removing the orbital contribution by a planar ramp.

5.2. Parameter Estimation

For the estimation of absolute orbit errors, a least squares approach similar to geodetic network adjustment is proposed, in which azimuth and range components are jointly estimated. Whereas a sequential adjustment has been preferred in (BÄHR AND HANSSEN, 2012), both a sequential and a closed approach are described in the following. The respective estimators are derived in detail in sections 5.2.2 and 5.2.3. The derivations are succeeded by detailed discussions of parameterisation, stochastic modelling and datum definition.

5.2.1. Hierarchical Organisation

Consider a network of n interferograms that are linear combinations of m images. An $n_{\varphi,k} \times 1$ vector φ_k of phase observations can be obtained from any interferogram k, and the final objective of network adjustment is the estimation of two orbit error parameters per acquisition or image, respectively. The latter are denoted in the style of the baseline error parameters introduced in section 4.3.1 and subsumed by the $2m \times 1$ parameter vector:

$$\mathbf{x} = \begin{pmatrix} \delta \dot{x}_{\|,1} \\ \delta x_{\perp,1} \\ \vdots \\ \delta \dot{x}_{\|,m} \\ \delta x_{\perp,m} \end{pmatrix}. \tag{5.1}$$

In the following, a closed and a sequential approach are distinguished (see figure 5.1), denoting specialised symbols by a subscript "s" for "sequential" or a subscript "c" for "closed", respectively. The functional model of a closed adjustment reads:

$$\mathrm{E}\left\{ \begin{pmatrix} \varphi_1 \\ \vdots \\ \varphi_n \end{pmatrix} \right\} = \mathbf{A}_c \mathbf{x}_c . \tag{5.2}$$

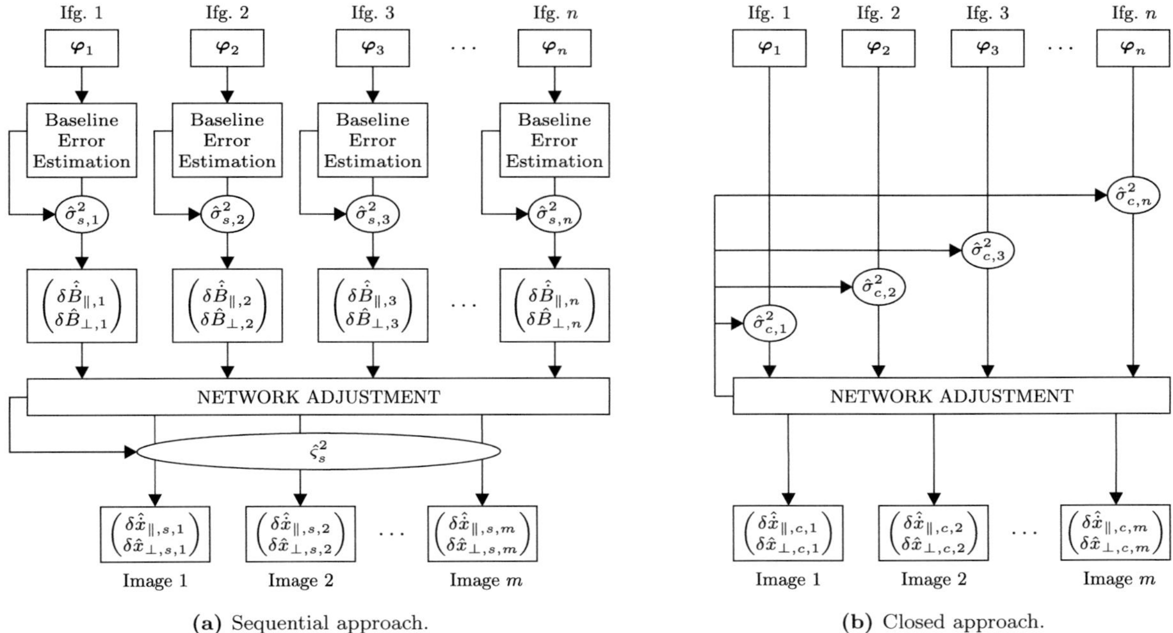

(a) Sequential approach.

(b) Closed approach.

Figure 5.1.: Two approaches for network adjustment of orbit errors are considered. **(a)** In the sequential approach, baseline errors $(\delta\dot{B}_\parallel, \delta B_\perp)$ are estimated as an intermediate result with individually estimated variance factors $\hat{\sigma}^2_{s,k}$. Network adjustment of absolute orbit errors $(\delta\dot{x}_\parallel, \delta x_\perp)$ goes along with a scaling of their covariance matrix by a global factor $\hat{\varsigma}_s$. **(b)** The closed adjustment deduces image-wise orbit errors directly from the interferometric phase. Individual variance components $\sigma^2_{c,k}$ are rigorously estimated.

The functional model of the sequential approach is split into two parts. Baseline errors $\mathbf{b}_{\theta,k} = \left(\delta\dot{B}_{\parallel,k}\ \delta B_{\perp,k}\right)^T$ are estimated from the individual interferograms in a first step. Subsequently, acquisition-specific orbit errors are adjusted to the individual baseline errors:

$$
\begin{aligned}
\mathrm{E}\{\boldsymbol{\varphi}_1\} &= \bar{\mathbf{A}}_{b,1}\mathbf{T}^T\mathbf{b}_{\theta,1} \\
&\vdots \\
\mathrm{E}\{\boldsymbol{\varphi}_n\} &= \bar{\mathbf{A}}_{b,n}\mathbf{T}^T\mathbf{b}_{\theta,n}
\end{aligned}
\qquad\longrightarrow\qquad
\mathrm{E}\left\{\begin{pmatrix}\hat{\mathbf{b}}_{\theta,1} \\ \vdots \\ \hat{\mathbf{b}}_{\theta,n}\end{pmatrix}\right\} = \mathbf{A}_s\mathbf{x}_s .
\tag{5.3}
$$

The closed approach is definitely the more rigorous formulation, because it accounts for different look directions $\vec{r}_\mathrm{M}$ and $\vec{r}_\mathrm{S}$ of master and slave, respectively, as it has been proposed by Kohlhase et al. (2003). The respective coefficients corresponding to errors in the master and the slave orbit do not need to be averaged as in eq. (4.7). The sequential approach on the other hand involves a bias due to averaging. But as the look directions of master and slave are almost collinear in spaceborne SAR, this bias is small enough to be negligible.

A practical advantage of the sequential approach is that it is well modularisable, because only the n two-element baseline error vectors and their covariance matrices are required as input data for the second step. But this is only a weak advantage, since the closed approach can also be modularised by stacking individual contributions to the normal equations.

Whereas both approaches involve rigorous variance propagation, an unbiased estimation of weights for the contributions of individual interferograms is only possible in a closed adjustment. In this one-step algorithm, individual variance components $\sigma^2_{c,k}$ are estimated from the level of consistency between observations and final estimates $\hat{\mathbf{x}}$ (see figure 5.1). In the two-step sequential approach, the estimation of

$\sigma_{s,k}^2$ can only rely on the consistency with the intermediate baseline error estimates $\hat{\mathbf{b}}_{\theta,k} = \left(\delta\hat{B}_{\|,k} \ \delta\hat{B}_{\perp,k} \right)^T$. In the second step, the overall variance level is adjusted by a global variance factor ς_s^2. Such a global scaling factor is not necessary in the closed approach, in which its estimate would always be 1 due to prior variance component estimation.

The performance of both approaches is evaluated in detail in chapter 6. Conclusions are drawn in section 6.2.4.

5.2.2. Adjustment in a Sequential Approach

The sequential approach of network adjustment combines individual estimation of baseline errors $\mathbf{b}_{\theta,k} = \left(\delta\dot{B}_{\|,k} \ \delta B_{\perp,k} \right)^T$, $k = 1 \ldots n$, as described in section 4.3 with a subsequent adjustment of misclosures between baseline errors obtained from linearly dependent interferometric combinations. The result is a consistent set of orbit error estimates $\hat{\mathbf{x}}_{s,i} = \left(\delta\hat{\dot{x}}_{\|,s,i} \ \delta\hat{x}_{\perp,s,i} \right)^T$, $i = 1 \ldots m$.

If interferogram k is formed from master acquisition i and slave acquisition j ($i, j \in \{1 \ldots m\}$), the functional model $\mathrm{E}\{\mathbf{l}_s\} = \mathbf{A}_s \mathbf{x}_s$ reads:

$$\mathrm{E}\left\{ \begin{pmatrix} \vdots \\ \hat{\mathbf{b}}_{\theta,k} \\ \vdots \end{pmatrix} \right\} = \begin{pmatrix} \cdots & \cdots & \cdots & \cdots & \cdots \\ \mathbf{0} & -\mathbf{I}_2 & \mathbf{0} & \mathbf{I}_2 & \mathbf{0} \\ \cdots & \cdots & \cdots & \cdots & \cdots \end{pmatrix} \begin{pmatrix} \vdots \\ \mathbf{x}_{s,i} \\ \vdots \\ \mathbf{x}_{s,j} \\ \vdots \end{pmatrix} . \tag{5.4}$$

$\mathbf{I}_2$ is a 2×2 identity matrix. If the baseline errors have been estimated with the least squares estimator from section 4.3.1, the stochastic model is based on the individual covariance matrices (4.19):

$$\mathrm{D}\{\mathbf{l}_s\} = \varsigma_s^2 \, \mathbf{P}_s^{-1} = \varsigma_s^2 \begin{pmatrix} \hat{\sigma}_{s,1}^2 \mathbf{Q}_{\theta,1} & & \mathbf{0} \\ & \ddots & \\ \mathbf{0} & & \hat{\sigma}_{s,n}^2 \mathbf{Q}_{\theta,n} \end{pmatrix} . \tag{5.5}$$

If, however, the baseline errors have been obtained with the gridsearch estimator from section 4.3.2, no covariance information is available. In this case, the most evident choice is to assume equal weights for all interferograms and define the mutual weighting of $\delta\hat{\dot{B}}_\|$ and $\delta\hat{B}_\perp$ by the fringe equivalents from eq. (3.17):

$$\mathrm{D}\{\mathbf{l}_s\} = \varsigma_s^2 \, \mathbf{P}_s^{-1} = \varsigma_s^2 \, \mathbf{I}_n \otimes \begin{pmatrix} (\delta\dot{B}_{\|,2\pi})^2 & 0 \\ 0 & (\delta B_{\perp,2\pi})^2 \end{pmatrix} , \tag{5.6}$$

where $\otimes$ is the Kronecker product. As the estimation of absolute orbit errors $\mathbf{x}_{s,k}$ requires a datum definition, two zero-mean conditions are introduced:

$$\sum_{k=1}^{m} \hat{\mathbf{x}}_{s,k} = \sum_{k=1}^{m} \begin{pmatrix} \delta\hat{\dot{x}}_{\|,s,k} \\ \delta\hat{x}_{\perp,s,k} \end{pmatrix} = \mathbf{0} . \tag{5.7}$$

The regularised normal equation system yielding a minimum-norm solution reads:

$$\begin{pmatrix} \mathbf{A}_s^T \mathbf{P}_s \mathbf{A}_s & \mathbf{G}^T \\ \mathbf{G} & \mathbf{0} \end{pmatrix} \begin{pmatrix} \hat{\mathbf{x}}_s \\ \mathbf{k}_s \end{pmatrix} = \begin{pmatrix} \mathbf{A}_s^T \mathbf{P}_s \mathbf{l}_s \\ \mathbf{0} \end{pmatrix} \tag{5.8}$$

with a $2 \times 2m$ matrix:

$$\mathbf{G} = \begin{pmatrix} \mathbf{I}_2 & \mathbf{I}_2 & \cdots & \mathbf{I}_2 \end{pmatrix} \tag{5.9}$$

and Lagrangian multipliers $\mathbf{k}_s$. Orbit error estimates $\hat{\mathbf{x}}_s$ are obtained from solving eq. (5.8). The corresponding covariance matrix $D\{\hat{\mathbf{x}}_s\} = \varsigma_s^2 \mathbf{Q}_{\hat{\mathbf{x}}_s \hat{\mathbf{x}}_s}$ is a submatrix of:

$$D\left\{ \begin{pmatrix} \hat{\mathbf{x}}_s \\ \mathbf{k}_s \end{pmatrix} \right\} = \varsigma_s^2 \begin{pmatrix} \mathbf{Q}_{\hat{\mathbf{x}}_s \hat{\mathbf{x}}_s} & \mathbf{Q}_{\hat{\mathbf{x}}_s \mathbf{k}_s} \\ \mathbf{Q}_{\mathbf{k}_s \hat{\mathbf{x}}_s} & \mathbf{Q}_{\mathbf{k}_s \mathbf{k}_s} \end{pmatrix} = \varsigma_s^2 \begin{pmatrix} \mathbf{A}_s^T \mathbf{P}_s \mathbf{A}_s & \mathbf{G}^T \\ \mathbf{G} & \mathbf{0} \end{pmatrix}^{-1} \tag{5.10}$$

The variance factor estimate:

$$\hat{\varsigma}_s^2 = \frac{\mathbf{v}_s^T \mathbf{P}_s \mathbf{v}_s}{2(n - m + 1)} \tag{5.11}$$

is obtained from the predicted corrections $\mathbf{v}_s = \begin{pmatrix} \cdots & v_{B_{\parallel},k} & v_{B_{\perp},k} & \cdots \end{pmatrix}^T$:

$$\mathbf{v}_s = \mathbf{A}_s \hat{\mathbf{x}}_s - \mathbf{l}_s \ . \tag{5.12}$$

5.2.3. Adjustment in a Closed Approach

The closed approach of network adjustment is a generalisation of the baseline error estimator derived in section 4.3.1. Alternatively, the baseline error estimator can also be considered a specialisation of closed network adjustment with $n = 1$ interferogram combining $m = 2$ images.

With the observation vector $\mathbf{l}_c = \begin{pmatrix} \varphi_1 & \cdots & \varphi_n \end{pmatrix}^T$, the closed functional model comprises $\sum_{k=1}^{n} n_{\varphi,k}$ observation equations (4.6):

$$\mathrm{E}\{\mathbf{l}_c\} = \mathbf{A}_c \begin{pmatrix} \delta x_{h,c,1} \\ \delta \dot{x}_{h,c,1} \\ \delta x_{v,c,1} \\ \delta \dot{x}_{v,c,1} \\ \delta x_{h,c,2} \\ \vdots \\ \delta \dot{x}_{v,c,m} \end{pmatrix} + \mathbf{A}_\varphi \begin{pmatrix} \varphi_{0,1} \\ \vdots \\ \varphi_{0,n} \end{pmatrix} \ . \tag{5.13}$$

The content of the first design matrix $\mathbf{A}_c$ can be inferred from eq. (4.6). $\mathbf{A}_\varphi$ is a block-diagonal matrix with $n_{\varphi,k} \times 1$ vectors of ones:

$$\mathbf{A}_\varphi = \begin{pmatrix} \mathbf{1} & & \mathbf{0} \\ & \ddots & \\ \mathbf{0} & & \mathbf{1} \end{pmatrix} \ . \tag{5.14}$$

The stochastic model:

$$D\{\mathbf{l}_c\} = \mathbf{P}_c^{-1} = \begin{pmatrix} \sigma_{c,1}^2 \mathbf{Q}_{\varphi,1} & & \mathbf{0} \\ & \ddots & \\ \mathbf{0} & & \sigma_{c,n}^2 \mathbf{Q}_{\varphi,n} \end{pmatrix} \tag{5.15}$$

consists of n scalable cofactor matrices $\mathbf{Q}_{\varphi,k}$ (see section 4.6), where the variance components $\sigma_{c,k}^2$ are a priori assumed to 1. Eliminating the phase offset parameters $\varphi_{0,k}$ from the functional model (5.13), a reduced design matrix $\bar{\mathbf{A}}_c$ is obtained by analogy to eq. (4.11):

$$\bar{\mathbf{A}}_c = \left(\mathbf{I} - \mathbf{A}_\varphi (\mathbf{A}_\varphi^T \mathbf{P}_c \mathbf{A}_\varphi)^{-1} \mathbf{A}_\varphi^T \mathbf{P}_c\right)\mathbf{A}_c \ . \tag{5.16}$$

Similar to eq. (4.15), the parameter space is restricted to the well determined components $\delta \dot{x}_{\parallel,c,k}$ and $\delta x_{\perp,c,k}$ subsumed by $\mathbf{x}_c$:

$$\mathrm{E}\{\mathbf{l}_c\} = \bar{\mathbf{A}}_c \mathbf{T}_c^T \mathbf{x}_c \tag{5.17}$$

by means of the transformation matrix:

$$\mathbf{T}_c = \mathbf{I}_m \otimes \begin{pmatrix} 0 & \sin\theta_0 & 0 & -\cos\theta_0 \\ \cos\theta_0 & 0 & \sin\theta_0 & 0 \end{pmatrix} \ . \tag{5.18}$$

With zero-mean conditions similar to eq. (5.7):

$$\sum_{k=1}^{m} \hat{\mathbf{x}}_{c,k} = \sum_{k=1}^{m} \begin{pmatrix} \delta \hat{\dot{x}}_{\parallel,c,k} \\ \delta \hat{x}_{\perp,c,k} \end{pmatrix} = \mathbf{0} \ , \tag{5.19}$$

the regularised normal equation system reads:

$$\begin{pmatrix} \mathbf{T}_c \bar{\mathbf{A}}_c^T \mathbf{P}_c \bar{\mathbf{A}}_c \mathbf{T}_c^T & \mathbf{G}^T \\ \mathbf{G} & \mathbf{0} \end{pmatrix} \begin{pmatrix} \hat{\mathbf{x}}_c \\ \mathbf{k}_c \end{pmatrix} = \begin{pmatrix} \mathbf{T}_c \bar{\mathbf{A}}_c^T \mathbf{P}_c \mathbf{l}_c \\ \mathbf{0} \end{pmatrix} \tag{5.20}$$

with Lagrangian multipliers $\mathbf{k}_c$. The parameters $\hat{\mathbf{x}}_c$ are obtained from inverting this system, and their covariance matrix $\mathrm{D}\{\hat{\mathbf{x}}_c\} = \mathbf{Q}_{\hat{\mathbf{x}}_c \hat{\mathbf{x}}_c}$ is contained in the inverse normal equation matrix:

$$\mathrm{D}\left\{ \begin{pmatrix} \hat{\mathbf{x}}_c \\ \mathbf{k}_c \end{pmatrix} \right\} = \begin{pmatrix} \mathbf{Q}_{\hat{\mathbf{x}}_c \hat{\mathbf{x}}_c} & \mathbf{Q}_{\hat{\mathbf{x}}_c \mathbf{k}_c} \\ \mathbf{Q}_{\mathbf{k}_c \hat{\mathbf{x}}_c} & \mathbf{Q}_{\mathbf{k}_c \mathbf{k}_c} \end{pmatrix} = \begin{pmatrix} \mathbf{T}_c \bar{\mathbf{A}}_c^T \mathbf{P}_c \bar{\mathbf{A}}_c \mathbf{T}_c^T & \mathbf{G}^T \\ \mathbf{G} & \mathbf{0} \end{pmatrix}^{-1} \ . \tag{5.21}$$

Variance Component Estimation

The n scaling factors $\sigma_{c,1}^2$ through $\sigma_{c,n}^2$ are obtained by means of variance component estimation (KOCH, 1999, p. 225 et seqq.). Within this framework, a multitude of different approaches have been developed (AMIRI-SIMKOOEI, 2007, p. 21 et seqq.). As rigorous estimators usually involve a considerable computational load, a simplified method originally proposed by FÖRSTNER (1979) is preferred for the current application. Iterative application of FÖRSTNER's estimator yields unbiased variance components at the point of convergence, whereas convergence is not guaranteed. However, for decorrelated observation groups as in eq. (5.15), convergence is usually achieved quickly. The estimator reads (BÄHR ET AL., 2007, p. 11):

$$\hat{\sigma}_{c,k}^2 = \frac{\mathbf{v}_{c,k}^T \mathbf{P}_{c,k} \mathbf{v}_{c,k}}{n_{\varphi,k} - u_k} \ . \tag{5.22}$$

$\mathbf{P}_{c,k}$ is the kth diagonal block of the weight matrix $\mathbf{P}_c$, and $\mathbf{v}_{c,k}$ is the corresponding partition of the vector of corrections:

$$\mathbf{v}_c = \bar{\mathbf{A}}_c \mathbf{T}_c^T \hat{\mathbf{x}}_c - \mathbf{l}_c \ . \tag{5.23}$$

u_k is a number that quantifies the proportionate contribution of the kth interferogram to the determination of all parameters, satisfying $\sum_{k=1}^{n} u_k = 2m + n - 2$. This sum is constituted of two parameters $(\delta \dot{x}_{\parallel,c,i}, \delta x_{\perp,c,i})$ per acquisition, one phase offset $\varphi_{0,k}$ per interferogram less the datum defect of two.

BÄHR ET AL. (2007) derive the computation of u_k in a Gauß-Markov model, in which no parameters have been eliminated. Conferred to the present adjustment problem it is:

$$u_k = \text{tr} \left\{ \mathbf{Q}^*_{\hat{\mathbf{x}}_c \hat{\mathbf{x}}_c} \mathbf{A}^{*T}_{c,k} \mathbf{P}_{c,k} \mathbf{A}^*_{c,k} \right\} ,$$

(5.24)

where the asterisk refers to the model, in which the phase offsets $\varphi_{0,k}$ have not been eliminated:

$$\mathbf{A}^*_c = \begin{pmatrix} \mathbf{A}_c \mathbf{T}_c & \mathbf{A}_\varphi \end{pmatrix} , \qquad \begin{pmatrix} \mathbf{Q}^*_{\hat{\mathbf{x}}_c \hat{\mathbf{x}}_c} & \mathbf{Q}^*_{\hat{\mathbf{x}}_c \mathbf{k}_c} \\ \mathbf{Q}^*_{\mathbf{k}_c \hat{\mathbf{x}}_c} & \mathbf{Q}_{\mathbf{k}_c \mathbf{k}_c} \end{pmatrix} = \begin{pmatrix} \mathbf{A}^{*T}_c \mathbf{P}_c \mathbf{A}^*_c & \mathbf{G}^T \\ \mathbf{G} & \mathbf{0} \end{pmatrix}^{-1} .$$

(5.25)

$\mathbf{A}^*_{c,k}$ is the contribution of interferogram k to the design matrix $\mathbf{A}^*_c$. The equivalent formulation of eq. (5.24) in the model with eliminated phase offsets (5.17) reads:

$$u_k = \text{tr} \left\{ \mathbf{Q}_{\hat{\mathbf{x}}_c \hat{\mathbf{x}}_c} \mathbf{T}_c \bar{\mathbf{A}}^T_{c,k} \mathbf{P}_{c,k} \bar{\mathbf{A}}_{c,k} \mathbf{T}^T_c \right\} - \frac{\mathbf{1}^T \mathbf{P}_{c,k} \bar{\mathbf{A}}_{c,k} \mathbf{T}^T_c \mathbf{Q}_{\hat{\mathbf{x}}_c \hat{\mathbf{x}}_c} \mathbf{T}_c \bar{\mathbf{A}}^T_{c,k} \mathbf{P}_{c,k} \mathbf{1}}{\mathbf{1}^T \mathbf{P}_{c,k} \mathbf{1}} + 1 ,$$

(5.26)

where $\bar{\mathbf{A}}_{c,k}$ is the contribution of interferogram k to the reduced design matrix $\bar{\mathbf{A}}_c$.

It is evident that the closed approach is only practicable for the least squares estimator. Applying the gridsearch principle in a closed manner would imply searching a $2m$-dimensional parameter space, which is far beyond any computationally manageable complexity.

5.2.4. Homogenisation of the Parameterisation

The Frenet frame defined in eq. (3.2) differs slightly for the individual acquisitions of a scene, because the orbit trajectories are in general not rigorously parallel. Even though the variation of individual frames is not very significant, a rigorous adjustment requires a unified frame. Thus, a reference acquisition is arbitrarily chosen to define a common $(\vec{e}_h(t), \vec{e}_v(t))$-plane onto which the baselines of all individual acquisitions are projected. This plane is always orthogonal to the curved orbit trajectory of the reference acquisition.

To find the correct Frenet frame for the computation of the coefficients $a_{h,\mathrm{M}}$, $a_{h,\mathrm{S}}$, $a_{v,\mathrm{M}}$ and $a_{v,\mathrm{S}}$ in eq. (4.6), locations on the orbit trajectory of acquisition k have to be related to corresponding locations on the reference trajectory (see figure 5.2). The azimuth time specifying the location on the reference orbit from where the same spot on the surface is acquired as from $\vec{x}(t_{k,i})$ on the orbit of acquisition k is:

$$t_{\mathrm{ref},i} + \Delta t_{\mathrm{ref},k} = t_{k,i} - t_{k,1} + t_{\mathrm{ref},1} + \Delta t_{\mathrm{ref},k} .$$

(5.27)

$t_{\mathrm{ref},1}$ is the start time of the reference acquisition, $t_{k,1}$ is the start time of acquisition k, and $\Delta t_{\mathrm{ref},k}$ is a constant offset. To determine the time offsets $\Delta t_{\mathrm{ref},k}$ for all acquisitions, the relative offsets $\Delta t_{k,l}$ are computed for all pairings (k,l) with contributing interferograms. After least squares adjustment of misclosures, for instance between the acquisitions j, k and l:

$$\Delta t_{j,k} + \Delta t_{k,l} + \Delta t_{l,j} \overset{!}{=} 0 ,$$

(5.28)

$m - 1$ consistent offsets $\Delta t_{\mathrm{ref},k}$ are obtained $(j,k,l,\mathrm{ref} \in \{1 \ldots m\})$.

Recall that the decomposition of the baseline into parallel and perpendicular component (3.4) depends on the look angle θ. For the estimators in section 4.3, a mean look angle was deduced from the error ellipses of the baseline estimates. In the network approach, which involves interferograms with different

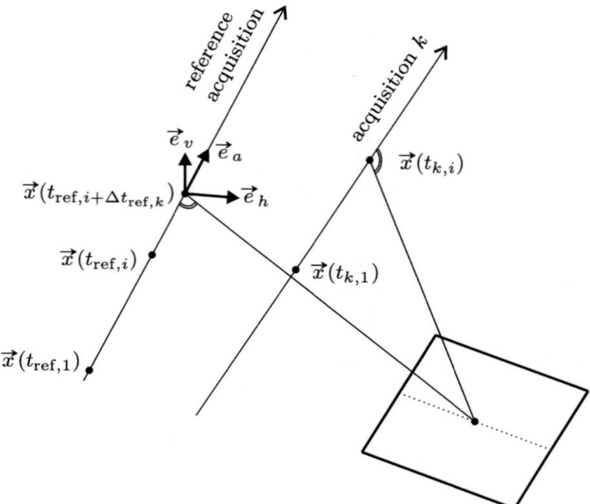

Figure 5.2.: Geometrical relation between the timescales of acquisition k and the reference acquisition that defines the Frenet frame. $t_{\mathrm{ref},1}$ and $t_{k,1}$ are the acquisition start times; $t_{\mathrm{ref},i}$ and $t_{k,i}$ refer to the respective ith pixel in azimuth.

acquisition geometries, θ_0 is heuristically averaged from the orientations of the baseline error ellipses of all n interferograms:

$$\theta_0 = \frac{1}{2n} \sum_{k=1}^{n} \left(\bar{\theta}_{0,k} + \bar{\theta}_{1,k} \right) . \tag{5.29}$$

$\bar{\theta}_{0,k}$ and $\bar{\theta}_{1,k}$ are obtained from eqs. (4.13) and (4.14), respectively.

Usually, the variation of individual Frenet frames is marginal, and the individually averaged look angles differ only on the $0.1°$ level. Whereas both effects have been considered within the scope of this thesis, an omission of homogenisation would probably not affect the adjustment results significantly.

5.2.5. Remarks on the Stochastic Model

The stochastic model of network adjustment has both a spatial and a temporal dimension. Besides the spatial correlation at different scales, which is mainly due to atmospheric turbulence and has been discussed in section 4.6, there is also some stochastic dependency between phase observations of different interferograms. Whereas this temporal correlation evidently comprises the effect of continuous deformation processes, it is rather dominated by algebraic correlation resulting from common images shared by different interferograms. Thus, almost all phase contributions are correlated between certain interferograms, because they are partly based on identical measurements of the undifferentiated phase ψ. The only component that is unaffected by algebraic correlation is the processing noise of interferogram formation, which is by the way the unique cause of network misclosures.

The stochastic models (5.5) and (5.15) completely neglect correlations between interferograms for the sake of simplicity. However, adequately considering these dependencies is a demanding challenge that can never be completed without compromises.

Modelling algebraic correlations would be straightforward if the stochastic properties of the phase could be adequately characterised by a unique covariance matrix $\mathbf{Q}_\varphi$ for all interferograms. Then, an appropriate stochastic model $D\{\mathbf{l}_c\}$ for the closed approach would be a block matrix that is assembled of sub-matrices $\mathbf{Q}_\varphi$ (scaled by integer coefficients 2, 0, 1 and -1) and regularised by an additional contribution of processing noise. However, the stochastic characteristics of individual interferograms are very different, and a uniform treatment would not adequately reflect the diversity of atmospheric states.

A most intuitive approach to generate a rigorous stochastic model would start from covariance matrices of the images and yield a global covariance matrix for all interferograms by variance propagation. However, it is not possible to obtain empirical covariance estimates for the undifferentiated phases ψ_i, since their expectation is unknown.

An alternative approach would be the empirical estimation of covariances between the phases φ_i of different interferograms. But as the degree of physical correlation does not depend on the temporal baseline B_T, a closed formulation of a spatio-temporal covariance function $C(r, B_T)$ is not feasible, and individual functions $C_{kl}(r)$ would have to be estimated for every pairing (k, l) of interferograms $(k, l = 1 \ldots n)$. However, a global covariance matrix constructed from multiple independently estimated covariance functions is not guaranteed to be positive definite.

Even if it was possible to find an adequate model for a covariance matrix of all phase observations from all interferograms, this matrix would have the dimension $(n \sum_{i=1}^{n} n_{\varphi,i}) \times (n \sum_{i=1}^{n} n_{\varphi,i})$, and the computational complexity of a dedicated estimator would increase dramatically with the number of interferograms, making the estimation impracticable. Consequently, neglecting mathematical correlations in the stochastic model is considered a necessary compromise.

Least squares network adjustment with the proposed stochastic model involves a weighting of baseline error estimates by interferogram-specific variance factors $\sigma_{s,k}^2$ or $\sigma_{c,k}^2$, respectively. These factors allocate higher weights to interferograms whose residual phase patterns closely resemble orbital error signals. Thus, interferograms with strong atmospheric signals that do not fit into the orbital signal space (see figure 4.8) are downweighted, mitigating their influence on the estimates. Such a weighting scheme is not provided in combination with the gridsearch estimator, for which the unweighted stochastic model (5.6) applies.

5.2.6. Datum Definition and Regularisation

As only relative orbit errors are estimable from the interferometric phase, network adjustment of two orbit error parameters per acquisition has a datum defect of two. The simplest approach for its elimination would be to consider the orbit of one acquisition deterministic. However, this would necessitate the arbitrary accentuation of a specific acquisition, whereas there is no physical motivation to consider the orbit of this one acquisition more accurate than the orbits of others.

For the zero-mean conditions (5.7) and (5.19), respectively, there is no physical justification either, but they involve less arbitrary choices than any other datum definition. They are expected to yield minimally biased estimates if the number of images is large and orbit errors are random. Their introduction minimises the norm of the solution vector ($\hat{\mathbf{x}}_s$ or $\hat{\mathbf{x}}_c$, respectively) as well a as the trace of the corresponding cofactor matrix ($\mathbf{Q}_{\hat{\mathbf{x}}_s \hat{\mathbf{x}}_s}$ or $\mathbf{Q}_{\hat{\mathbf{x}}_c \hat{\mathbf{x}}_c}$, respectively).

Several approaches to regularise the normal equation system can be distinguished. In eqs. (5.8) and (5.20), respectively, full rank is procured by bordering the normal equation matrix and extending the parameter set by Lagrangian multipliers (KOCH, 1999, pp. 185 et seqq.; JÄGER ET AL., 2006, pp. 243 et seqq.; NIEMEIER, 2008, pp. 27 et seqq.). An equivalent result could alternatively be obtained by means of the pseudo-inverse or a singular value decomposition of the normal equation matrix $\mathbf{A}_s^T \mathbf{P}_s \mathbf{A}_s$ or $\mathbf{T}_c \bar{\mathbf{A}}_c^T \mathbf{P}_c \bar{\mathbf{A}}_c \mathbf{T}_c^T$, respectively. The here proposed approach has its strengths in the enhanced flexibility of datum definition. In some cases it is desirable to consider only a subset of acquisitions for the minimum-norm condition, yielding a so-called partial trace minimisation. For instance, if the network is extended after an initial adjustment by new, recently acquired images, these should be disregarded in the datum conditions eq. (5.7) or (5.19), respectively, to avoid a systematic shift of all estimates. This is implemented by zeroing the corresponding coefficients in $\mathbf{G}$.

The potential of the regularisation approach is illustrated by the following example: For $m = 3$ images with orbit error parameters $\mathbf{x}_k = \left(\delta \dot{x}_{\parallel,k} \;\; \delta x_{\perp,k} \right)^T$, $k = 1 \ldots 3$, the zero-mean-conditions read $\sum_{k=1}^{3} \hat{\mathbf{x}}_k = \mathbf{0}$, and $\mathbf{G} = \left(\mathbf{I}_2 \; \mathbf{I}_2 \; \mathbf{I}_2 \right)$. If a fourth image with a large orbit error is added to the network later and the adjustment is re-performed with zero-mean conditions comprising all four images (i. e., $\sum_{k=1}^{4} \hat{\mathbf{x}}_k = \mathbf{0}$, $\mathbf{G} = \left(\mathbf{I}_2 \; \mathbf{I}_2 \; \mathbf{I}_2 \; \mathbf{I}_2 \right)$), the orbit error estimates would experience a significant shift also for the initial three images, even if all misclosures are zero. This effect can be avoided by excluding the fourth image from the zero-mean condition: $\sum_{k=1}^{3} \hat{\mathbf{x}}_k = \mathbf{0}$. Then, $\mathbf{G} = \left(\mathbf{I}_2 \; \mathbf{I}_2 \; \mathbf{I}_2 \; \mathbf{0} \right)$.

5.3. Outlier Detection

Geodetic network adjustment theory provides a powerful framework for quality control. Measures of internal reliability assess the sensitivity of a network of measured distances, horizontal directions, vertical angles and/or GPS observations with respect to outliers in particular observations (JÄGER ET AL., 2006). Under the hypothesis that an observation or a group of observations is affected by a gross error, the amount of error can be estimated, statistically tested and eliminated where appropriate.

Conferring these tools to the estimation of InSAR orbit errors requires some adaptions. The primary objective is to detect multi-pixel unwrapping errors of the least squares estimator, which cannot be attributed to either the $\delta \dot{B}_{\parallel}$- or the $\delta B_{\perp}$-component, as unwrapping is a two-dimensional process. But even if explicit unwrapping is circumvented by the gridsearch estimator, outliers may result from unreliable baseline error estimates due to ambiguous global maxima in the parameter space. In the context of the closed approach, it would mathematically be feasible to test suspect patches of pixels on common unwrapping errors, but this is not practical in view of the exorbitant high number of potential groupings of individual phase observations.

Consequently, the most promising strategy is to consider the contribution of an interferogram as a whole. Outlier assessment of the contribution of interferogram k is based on the hypothesis that the contributions of all interferograms except the kth one are free of blunders. In case of more than one contaminated interferogram, successful detection may still be accomplishable by iterative data snooping (see section 4.5.2), which may also fail in presence of too many outliers masking each other.

5.3.1. Sequential Approach

Considering eq. (5.4) the null hypothesis, the alternative hypothesis admitting a two-parametric bias $\boldsymbol{\nabla}_{s,k} = \left(\nabla \dot{B}_{\parallel,s,k} \;\; \nabla B_{\perp,s,k} \right)^T$ in the contribution of the kth interferogram reads:

$$E\{l_s\} = \mathbf{A}_s \mathbf{x}_s + \begin{pmatrix} \vdots \\ \mathbf{0} \\ \mathbf{I}_2 \\ \mathbf{0} \\ \vdots \end{pmatrix} \boldsymbol{\nabla}_{s,k} \, . \tag{5.30}$$

Assuming thus that interferogram k is the only interferogram that contributes biased estimates to the network, the best estimate for $\mathbf{\nabla}_{s,k}$ is (JÄGER ET AL., 2006, p. 109):

$$\hat{\mathbf{\nabla}}_{s,k} = -\left(\mathbf{P}_{s,k} - \mathbf{P}_{s,k}\mathbf{A}_{s,k}\mathbf{Q}_{\hat{\mathbf{x}}_s\hat{\mathbf{x}}_s}\mathbf{A}_{s,k}^T\mathbf{P}_{s,k}\right)^{-1}\mathbf{P}_{s,k}\mathbf{v}_{s,k} \,, \tag{5.31}$$

where $\mathbf{A}_{s,k}$ is the contribution of the kth interferogram to the design matrix $\mathbf{A}_s$, $\mathbf{P}_{s,k}$ is the kth diagonal block of $\mathbf{P}_s$, and $\mathbf{v}_{s,k} = \mathbf{A}_{s,k}\hat{\mathbf{x}}_s - \hat{\mathbf{b}}_{\theta,k}$ comprises the corresponding two elements of $\mathbf{v}_s$. The significance of $\hat{\mathbf{\nabla}}_{s,k}$ can be evaluated by a generalised outlier test (4.26) for two parameters (JÄGER ET AL., 2006, p. 191):

$$T_{s,k} = -\frac{\mathbf{v}_{s,k}^T\mathbf{P}_{s,k}\hat{\mathbf{\nabla}}_{s,k}}{2\hat{\varsigma}_s^2} \sim F_{2,2(n-m)} \,, \quad n > m \,. \tag{5.32}$$

By analogy to the one-parametric t-test in section 4.5.2, the variance factor estimate $\hat{\varsigma}_s^2$ from eq. (5.11) would be biased in presence of a blunder in the kth interferogram. Hence, its contribution is removed from the sum of squared corrections, yielding the adapted estimate:

$$\bar{\varsigma}_s^2 = \frac{\mathbf{v}_s^T\mathbf{P}_s\mathbf{v}_s + \mathbf{v}_{s,k}^T\mathbf{P}_{s,k}\hat{\mathbf{\nabla}}_{s,k}}{2(n-m)} \,. \tag{5.33}$$

For iterative outlier detection (data snooping), the following procedure is proposed: If $T_{s,k}$ exceeds both a dedicated threshold deduced from the Fisher distribution and any other $T_{s,j}$ with $j \in \{1,\ldots,n\}\backslash k$, it is checked in the first place if there is an unwrapping error in interferogram k that can be corrected manually. Otherwise, its contribution is rejected, and the procedure is repeated until all test statistics fall below the threshold. Rejection must not be pursued too extensively, guaranteeing that the contribution of every interferogram is controlled by at least one linear combination of other interferograms in the network.

5.3.2. Closed Approach

The formulation of an equivalent methodology in the closed approach is not straightforward. By analogy, the biases of the contribution from interferogram k could likewise be modelled by two nuisance parameters $\mathbf{\nabla}_{c,k} = \left(\nabla\dot{B}_{\parallel,c,k} \; \nabla B_{\perp,c,k}\right)^T$. Based on the null hypothesis (5.13), the alternative hypothesis would read:

$$\mathrm{E}\{\mathbf{l}_c\} = \bar{\mathbf{A}}_c\mathbf{T}_c^T\mathbf{x}_c + \begin{pmatrix} \vdots \\ \mathbf{0} \\ \mathbf{B}_k \\ \mathbf{0} \\ \vdots \end{pmatrix} \mathbf{\nabla}_{c,k} \tag{5.34}$$

with:

$$\mathbf{B}_k = \frac{1}{2}\left(\bar{\mathbf{A}}_{c,kj} - \bar{\mathbf{A}}_{c,ki}\right)\mathbf{T}^T \,. \tag{5.35}$$

$\bar{\mathbf{A}}_{c,ki}$ is the block of $\bar{\mathbf{A}}_c$ that corresponds to interferogram k and its master acquisition i, and $\bar{\mathbf{A}}_{c,kj}$ is the block corresponding to interferogram k and its slave acquisition j. An estimate:

$$\hat{\mathbf{\nabla}}_{c,k} = -\left(\mathbf{B}_k^T\mathbf{P}_{c,k}\mathbf{Q}_{\mathbf{v}_{c,k}\mathbf{v}_{c,k}}\mathbf{P}_{c,k}\mathbf{B}_k\right)^{-1}\mathbf{B}_k^T\mathbf{P}_{c,k}\mathbf{v}_{c,k} \tag{5.36}$$

could be obtained by analogy to the sequential approach. Its computation involves the kth diagonal block $\mathbf{Q}_{\mathbf{v}_{c,k}\mathbf{v}_{c,k}}$ of the cofactor matrix $\mathbf{Q}_{\mathbf{v}_c\mathbf{v}_c}$ of the predicted corrections $\mathbf{v}_c$, which is defined in the model (5.25) without eliminated phase offsets:

$$\mathbf{Q}_{\mathbf{v}_c\mathbf{v}_c} = \mathbf{P}_c^{-1} - \mathbf{A}_c^*\mathbf{Q}_{\hat{\mathbf{x}}_c\hat{\mathbf{x}}_c}^*\mathbf{A}_c^{*T} \,. \tag{5.37}$$

To establish a test statistic for $\hat{\bar{\nabla}}_{c,k}$, it has to be considered that the kth variance component $\hat{\sigma}^2_{c,k}$ is estimated from the very same subset of observations (i. e., $\mathbf{l}_{c,k} = \boldsymbol{\varphi}_k$) that is suspected to be contaminated by outliers in eq. (5.34). Hence, an unwrapping error in interferogram k would not only be absorbed by the dedicated nuisance parameters but also distort the corresponding variance component. However, unbiased variance components are indispensable for a statistically rigorous outlier test. In contrast to the sequential approach, where only a global variance factor ς^2_s is estimated, it is not feasible to remove the contribution of the suspect observations from the variance component estimate $\hat{\sigma}^2_{c,k}$ by analogy to eq. (5.33). As all contributing observations $\boldsymbol{\varphi}_k$ are considered potential outliers, there would be no assumedly correct observations left to rely on.

Realising that standard outlier statistics cannot be applied with statistical rigour here, the model (5.34) is dropped, and a more pragmatic approach is proposed: The test statistic from the sequential approach is adapted to the closed framework by simply replacing $\mathbf{x}_s$ by $\mathbf{x}_c$ in eq. (5.30):

$$
\mathrm{E}\{\mathbf{l}_s\} = \mathbf{A}_s \mathbf{x}_c + \begin{pmatrix} \vdots \\ \mathbf{0} \\ \mathbf{I}_2 \\ \mathbf{0} \\ \vdots \end{pmatrix} \bar{\mathbf{\nabla}}_{c,k} \; . \tag{5.38}
$$

Thus, the estimated bias:

$$
\hat{\bar{\mathbf{\nabla}}}_{c,k} = - \left(\mathbf{P}_{s,k} - \mathbf{P}_{s,k} \mathbf{A}_{s,k} \mathbf{Q}_{\hat{\mathbf{x}}_c \hat{\mathbf{x}}_c} \mathbf{A}^T_{s,k} \mathbf{P}_{s,k} \right)^{-1} \mathbf{P}_{s,k} \bar{\mathbf{v}}_{c,k} \tag{5.39}
$$

is computed from the modified corrections:

$$
\bar{\mathbf{v}}_c = \mathbf{A}_s \hat{\mathbf{x}}_c - \mathbf{l}_s \; . \tag{5.40}
$$

The corresponding test statistic reads:

$$
T_{c,k} = - \frac{\bar{\mathbf{v}}^T_{c,k} \mathbf{P}_{s,k} \hat{\bar{\mathbf{\nabla}}}_{c,k}}{2 \varsigma^2_c} \sim F_{2,2(n-m)} \tag{5.41}
$$

with:

$$
\varsigma^2_c = \frac{\bar{\mathbf{v}}^T_c \mathbf{P}_s \bar{\mathbf{v}}_c + \bar{\mathbf{v}}^T_{c,k} \mathbf{P}_{s,k} \hat{\bar{\mathbf{\nabla}}}_{c,k}}{2(n-m)} \; . \tag{5.42}
$$

Even though this test is not statistically rigorous, it turns out to be a good numerical approximation of the test $T_{s,k}$ that has been substantiatedly derived for the sequential approach. Whereas a good performance cannot be guaranteed in general, it is exemplarily demonstrated for a test case in section 6.3.

6. Application of Orbit Error Estimation

The proposed approach for estimating orbit errors has been tested on a set of $m = 31$ Envisat acquisitions from a scene in Western Australia (track 203, frame 4221) between December 2003 and April 2008. The region has a semi-arid climate, the land use being dominated by dryland cropping and some salt lakes. These conditions go along with a good interferometric coherence, which was the reason to choose this test area. Besides, no measurable ground deformation is expected for the period of data coverage. A network of $n = 163$ interferograms has been set up with a maximum perpendicular baseline of 743 m and a maximum temporal baseline of 560 days (see figure 6.1). It was aimed to include as many interferograms as possible, the only requirement being that unwrapping is reliably feasible (see figure 6.2 for four sample interferograms).

InSAR processing has been performed with the *Delft Object-Oriented Radar Interferometric Software* (DORIS; KAMPES ET AL., 2004) using precise orbits of the French *Centre National d'Etudes Spatiales* (CNES). Topographic height variations, which are less than 200 m, have been accounted for with a 3" DEM product from the *Shuttle Radar Topography Mission* (SRTM). To maximise coherence, all interferograms have been multilooked by a factor 25 in azimuth and 5 in range, yielding pixels of approximately 100×100 m^2 size. Adaptive phase filtering (GOLDSTEIN AND WERNER, 1998) has been applied to facilitate unwrapping, for which the *Statistical-Cost Network-Flow Algorithm for Phase Unwrapping* (SNAPHU; CHEN AND ZEBKER, 2001) has been used.

In the following, several variations of the proposed network approach are compared: least squares versus gridsearch estimator, sequential versus closed adjustment and uncorrelated versus correlated observa-

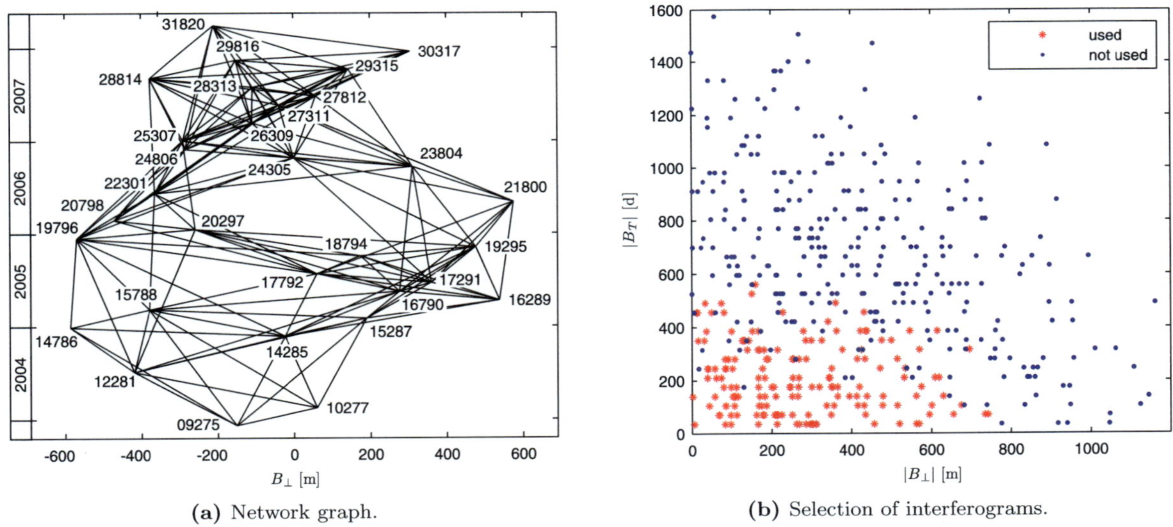

(a) Network graph.   (b) Selection of interferograms.

Figure 6.1.: Overview of the data set of $m = 31$ Envisat acquisitions (track 203, frame 4221) from Western Australia between December 2003 and April 2008. **(a)** $n = 163$ of the $m(m-1)/2 = 465$ possible interferometric combinations could be reliably unwrapped and have thus been considered for the network. **(b)** Most interferograms with a temporal baseline B_T above 500 days and a perpendicular baseline $B_\perp$ above 700 m have been disregarded.

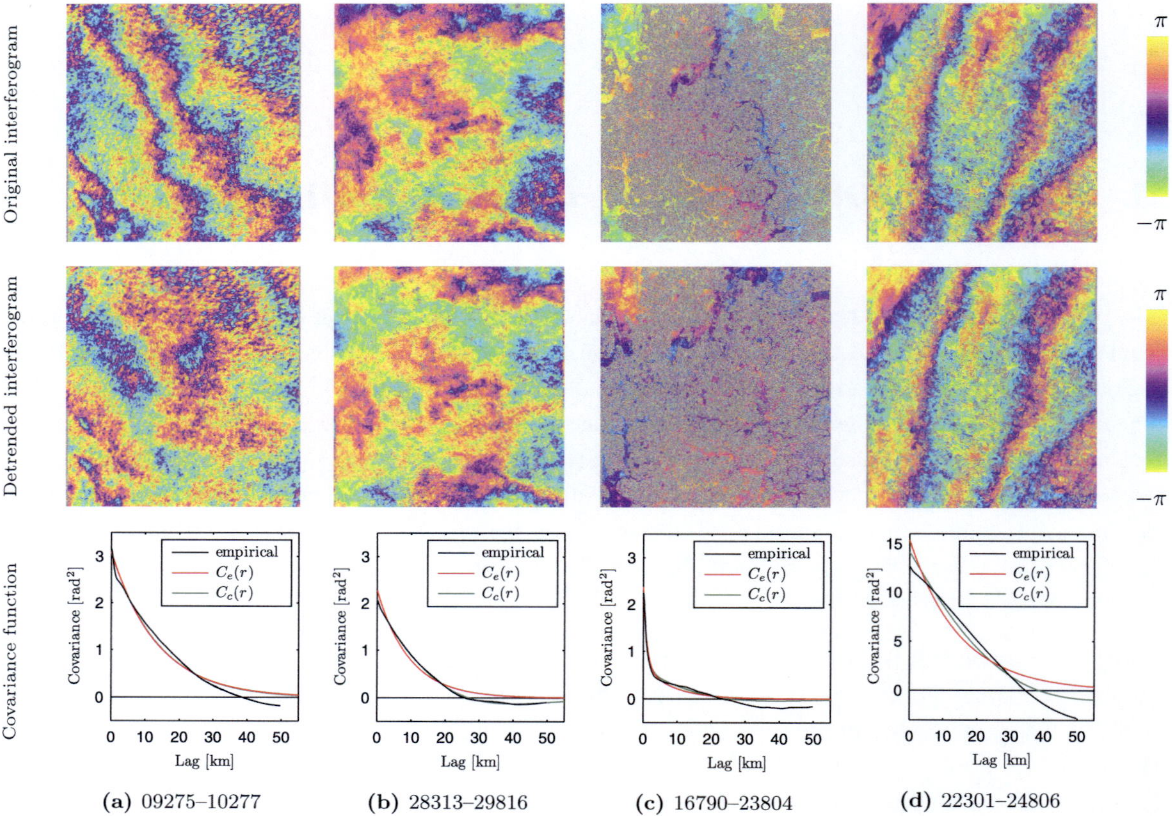

Figure 6.2.: Sample interferograms from the Envisat data set and their associated covariance functions. The displayed interferograms cover the whole scene of 100×100 km^2 and have been corrected for the reference phase with respect to the DEM. **(a)** The empirical covariance function of this interferogram is well approximable by the proposed covariance models ($B_\perp = 211$ m, $B_T = 70$ d). **(b)** In some cases, the double exponential model with cosine $C_c(r)$ involves a very small improvement with respect to the model $C_e(r)$ without cosine ($B_\perp = -44$ m, $B_T = 105$ d). **(c)** In spite of its long temporal baseline and large decorrelated areas, this interferogram is still reliably unwrappable ($B_\perp = 34$ m, $B_T = 490$ d). **(d)** Some interferograms, especially those related to acquisition 24806, exhibit a large-scale nonlinear signal. Consequently, the stationarity assumption is violated, and the obtained covariance function is biased ($B_\perp = 80$ m, $B_T = 175$ d).

tions. Furthermore, the capability of data snooping is evaluated. After a thorough analysis of results, conclusions are drawn in section 6.4.

6.1. Stochastic Model

One of the aims of this investigation is to evaluate the adequacy of different stochastic models. In this context, adequacy does not only imply statistical rigorousness but also general applicability and a reasonable compromise between effort and benefit. To explore the potential of minimum effort, the simple model (4.29) is included into the considerations, because it requires no individual adaption and may still show an acceptable performance. It assumes unweighted and uncorrelated observations and can formally be represented by the trivial covariance function:

$$C_0(r) = \begin{cases} 1, & r = 0 \\ 0, & r > 0 \end{cases}.$$

(6.1)

The a priori variance $C_0(0)$ is conventionally set to one. This can be done, because the variance level is scaled a posteriori by the variance factor estimates $\hat{\sigma}^2_{s,k}$ or $\hat{\sigma}^2_{c,k}$, respectively.

It is evident that neglecting correlations is not statistically rigorous as long as a significant atmospheric signal is contained in the interferograms. As this signal is not considered in the functional model, it needs to be accommodated by the stochastic model. Modelling its stochastic properties by means of a covariance function involves individual adjustment of some parameters and also requires second order stationarity of the atmospheric signal. However, the latter condition is not satisfied for the present network, which contains several interferograms with atmospheric signals of spatial wavelengths exceeding the size of an interferogram. It is impossible to fully describe a stochastic process if the only available sample is smaller than the correlation length of a characteristic component. From the perspective of the interferogram as a cutout window, such a process has to be considered non-stationary. Hence, a generally applicable methodology and statistically rigorous modelling cannot be provided both at a time, and compromises have to be made. The here pursued approach is based on tolerating non-stationarity to a certain extent for the sake of generality.

Thus considering an interferogram a realisation of a second-order stationary process, a two-dimensional covariance function can be estimated by application of an inverse Fourier transformation to its power spectrum (HANSSEN, 2001, p. 137 et seq.):

$$C(\Delta\xi, \Delta\eta) = \mathcal{F}^{-1}\left\{\mathcal{F}\left\{\varphi(\xi,\eta)\right\}\left(\mathcal{F}\left\{\varphi(\xi,\eta)\right\}\right)^*\right\} , \tag{6.2}$$

where $\mathcal{F}\{\cdot\}$ is the Fourier transform operator and $(\cdot)^*$ is the complex conjugate. As an eventual orbital signal is attributed to the functional and not to the stochastic model, it is removed from the interferogram beforehand by subtraction of a linear trend, which is a sufficiently good approximation for this purpose. The power spectrum $\mathcal{F}\left\{\varphi\right\}\left(\mathcal{F}\left\{\varphi\right\}\right)^*$ is also low-pass filtered to avoid a bias due to decorrelation noise. A one-dimensional covariance function is finally obtained by circular averaging, assuming isotropy of the underlying signal (JÓNSSON, 2002, p. 24):

$$C(r) = \frac{1}{2\pi r} \int_0^{2\pi} C(r\cos\vartheta, r\sin\vartheta)\, d\vartheta . \tag{6.3}$$

Due to the sampling theorem, the obtained covariance function is only defined for spatial wavelengths of less than half the size of the interferogram. To make it applicable for longer lags, it is extrapolated by fitting an analytical model, which is required to be positive definite. Having evaluated some dedicated generic models (e. g., WEBSTER AND OLIVER, 2007) and linear combinations thereof, two approaches are taken into closer consideration, because they match well the empirical covariance functions (see figure 6.2):

$$C_e(r; c_1, c_2, a) = c_1 e^{-\frac{r}{1\mathrm{km}}} + c_2 e^{-\frac{r}{a}} \tag{6.4}$$

$$C_c(r; c_1, c_2, a_2, a_3) = c_1 e^{-\frac{r}{1\mathrm{km}}} + c_2 e^{-\frac{r}{a_2}} \cos\left(2\pi\frac{r}{a_3}\right) . \tag{6.5}$$

The empirical choice of 1 km as correlation length of the first exponential function can be motivated with the transition between different atmospheric scaling regimes at 2 km distance (HANSSEN, 2001, p. 142 et seq.); $e^{-r/(1\mathrm{km})}$ has decayed by 86 % at $r = 2$ km.

Variable characteristics of the atmospheric signals result in distinctly individual covariance functions with different approximation quality (see figure 6.2). It is interesting to note that the parameter c_1 is practically insignificant (i. e., $< 0.01\,\mathrm{rad}^2$) for a considerable portion of interferograms (50 % with $C_e(r)$, 28 % with $C_c(r)$) while it is very relevant for others (e. g., 16790–23804, see figure 6.2c). The second model $C_c(r)$ has one parameter more than the simpler model $C_e(r)$ and can thus accomodate a hole effect, i. e., negative covariances at a certain lag (see figure 6.2b). Although the resulting improvement is insignificant for most interferograms, both models are considered in the following in order to analyse the sensitivity of the estimation with respect to small changes in the stochastic model.

As the correlated atmospheric signal is superposed by uncorrelated noise, a nugget effect may be considerable, i. e., a discontinuity of the covariance function at $r = 0$. However, the noise is probably negligible in relation to the atmospheric contribution as only the most coherent pixels are picked as observations.

Nevertheless, all potential enhancements cannot compensate the major deficiency of the pursued approach: the violation of the stationarity assumption for some interferograms with nonlinear large-scale atmospheric signals (e. g., ifg. 22301–24806, see figure 6.2). The most seriously biased covariance functions are obtained for interferograms from image 24806, but rejecting this image from the network would not be a prospective solution when aiming at a generally applicable methodology. Furthermore, its rejection would not solve the problem completely, since there are also other interferograms with less pronounced but still significant nonlinear atmospheric trends.

6.2. Estimation Results

Both sequential and closed adjustment approach yield numerically almost identical estimates. Consequently, the results of sequential adjustment, which are presented in the following, can be considered representative for both approaches. A concise quantification of numerical differences between the two as well as some conceptual considerations are provided in section 6.2.4.

6.2.1. Criteria of Evaluation and Comparison

For the comparison of two approaches, suitable criteria need to be defined. Central subject of any comparison are the baseline error estimates $\hat{\mathbf{b}}_{\theta,k} = \left(\delta\hat{\dot{B}}_{\|,k} \ \delta\hat{B}_{\perp,k} \right)^T$ for single interferograms ($k = 1 \ldots n$, eqs. (4.18) and (4.22)) and the orbit error estimates $\hat{\mathbf{x}} = \left(\cdots \ \delta\hat{\dot{x}}_{\|,i} \ \delta\hat{x}_{\perp,i} \ \cdots \right)^T$ on the network level ($i = 1 \ldots m$, eqs. (5.8) and (5.20)). Of further interest are the baseline error corrections $\mathbf{v}_s = \left(\cdots \ v_{\dot{B}_{\|,k}} \ v_{B_{\perp,k}} \ \cdots \right)^T$ (eq. (5.12)) and the estimated biases $\hat{\boldsymbol{\nabla}}_k = \left(\hat{\nabla}\dot{B}_{\|,k} \ \hat{\nabla}B_{\perp,k} \right)^T$ (eqs. (5.31) and (5.39)).

As the baseline error components $\delta\dot{B}_\|$ and $\delta B_\perp$ have a physical meaning, their estimates can be compared individually. However, a combined comparison accounts for the two-dimensional nature of the orbital error signal in a more adequate way. Thus, orbit errors are complementarily quantified in terms of the total number of fringes that they induce into an interferogram:

$$\delta\hat{B}_{\mathrm{fr}} := \left| \frac{\delta\hat{\dot{B}}_\|}{\delta\dot{B}_{\|,2\pi}} \right| + \left| \frac{\delta\hat{B}_\perp}{\delta B_{\perp,2\pi}} \right| . \tag{6.6}$$

The conversion to fringes follows eq. (3.17), whereas $\delta\dot{B}_{\|,2\pi} = 1.7\ \mathrm{mm/s}$ and $\delta B_{\perp,2\pi} = 26\ \mathrm{cm}$ are applicable for the data set at hand. An analogous conversion can be applied to absolute orbit errors $\hat{\mathbf{x}}$, predicted baseline error corrections $\mathbf{v}_s$ and estimated biases $\hat{\boldsymbol{\nabla}}$, yielding:

$$\delta\hat{x}_{\mathrm{fr}} := \left| \frac{\delta\hat{\dot{x}}_\|}{\delta\dot{B}_{\|,2\pi}} \right| + \left| \frac{\delta\hat{x}_\perp}{\delta B_{\perp,2\pi}} \right| \tag{6.7}$$

$$v_{B_{\mathrm{fr}}} := \left| \frac{v_{\dot{B}_\|}}{\delta\dot{B}_{\|,2\pi}} \right| + \left| \frac{v_{B_\perp}}{\delta B_{\perp,2\pi}} \right| \tag{6.8}$$

$$\hat{\nabla}B_{\mathrm{fr}} := \left| \frac{\hat{\nabla}\dot{B}_\|}{\delta\dot{B}_{\|,2\pi}} \right| + \left| \frac{\hat{\nabla}B_\perp}{\delta B_{\perp,2\pi}} \right| . \tag{6.9}$$

Due to lack of independent reference data, it is impossible to explicitly validate the estimates. However, the model consistency can be evaluated by testing if the predicted corrections follow the expected distribution, i. e., $\mathbf{v} \sim N(\mathbf{0}, \mathrm{D}\{\mathbf{v}\})$. Comprehensive indicators for this consistency are the variance factor estimates $\hat{\sigma}_{s,k}$, $\hat{\sigma}_{c,k}$ and $\hat{\varsigma}_{s}$.

The absolute deviations between the estimates of an approach A and an approach B are quantified by the following measures:

$$\Delta \delta \hat{B}_{\mathrm{fr,A/B}} := \left| \frac{\delta \hat{B}_{\parallel,A} - \delta \hat{B}_{\parallel,B}}{\delta \dot{B}_{\parallel,2\pi}} \right| + \left| \frac{\delta \hat{B}_{\perp,A} - \delta \hat{B}_{\perp,B}}{\delta B_{\perp,2\pi}} \right| \tag{6.10}$$

$$\Delta \delta \hat{x}_{\mathrm{fr,A/B}} := \left| \frac{\delta \hat{x}_{\parallel,A} - \delta \hat{x}_{\parallel,B}}{\delta \dot{B}_{\parallel,2\pi}} \right| + \left| \frac{\delta \hat{x}_{\perp,A} - \delta \hat{x}_{\perp,B}}{\delta B_{\perp,2\pi}} \right| . \tag{6.11}$$

Analogous deviation measures can be defined for the baseline error corrections $\mathbf{v}_s$ and the estimated biases $\hat{\nabla}$.

6.2.2. Individual Baseline Error Estimates

Interferogram-wise baseline error estimation has been performed as described in section 4.3. Subdividing all interferograms into 1224 tiles of 30×30 pixels, only the most coherent pixel from each tile contributes to the estimation to guarantee a spatially homogeneous distribution of observations. Between 0 and 104 tiles have been disregarded due to lack of pixels with a coherence estimate above 0.25. Additionally, data snooping has been applied as described in section 4.5.2 to ensure that the observations are free of exceptionally large outliers, using a 0.1 % criterion for the two-sided t-test (4.26). To avoid an unbalanced distribution of observations, the number of rejected tiles has been limited to 30. Finally, depending on the individual interferograms, between 1116 and 1224 phase observations have been used to estimate baseline error parameters.

It can be seen from table 6.1 that the least squares baseline error estimates have a similar order of magnitude for the three covariance models. However, in view of the expectable orbit accuracy discussed in section 3.3.3, it cannot be fully explained by orbit errors that estimates for $\delta \hat{B}_{\perp}$ are larger than $16 \dots 17$ cm for more than 50 % of the interferograms. It is evident that large-scale variations of the atmospheric delay leak into the baseline error estimates. The plausibility of the atmospheric origin of these signals is supported by the bulge-like trend in interferogram 22301–24806 (see figure 6.2). It has a gradient of two fringes over half a scene (50 km), which cannot be caused by orbit errors as it does not match the corresponding signal space (see figure 4.8a).

The product of the interferogram-specific a priori variances $C(0)$ and the variance factor estimates $\hat{\sigma}_{s,k}^2$ is an estimate for the variance of the interferometric phase with removed orbital contribution (see figure 6.3a). The corresponding standard deviations are in the order of 1 rad for most interferograms, ranging up to 4.1 rad. For uncorrelated observations ($C_0(r)$), the highest values can be observed for interferograms containing image 24806, which is affected by a bulge-like atmospheric signal (see figure 6.2d).

Of special interest are the variance factor estimates $\hat{\sigma}_{s,k}^2$ for the covariance models that are based on the a priori estimated covariance functions $C_e(r)$ or $C_c(r)$, respectively (see figure 6.3b). If these functions adequately described the stochastic behaviour of the observations, $\mathrm{E}\{\hat{\sigma}_s^2\} = 1$ and $\hat{\sigma}_s^2 \sim F_{n_\varphi,k-3,\infty}$ would hold. The validity of the Fisher distribution can be tested in a global model test, which fails for most of the interferograms ($C_e(r)$: 90 %; $C_c(r)$: 88 %). This can either imply deficiencies in the functional and the stochastic model or indicate blunders in the observations. As the observations have been pre-screened

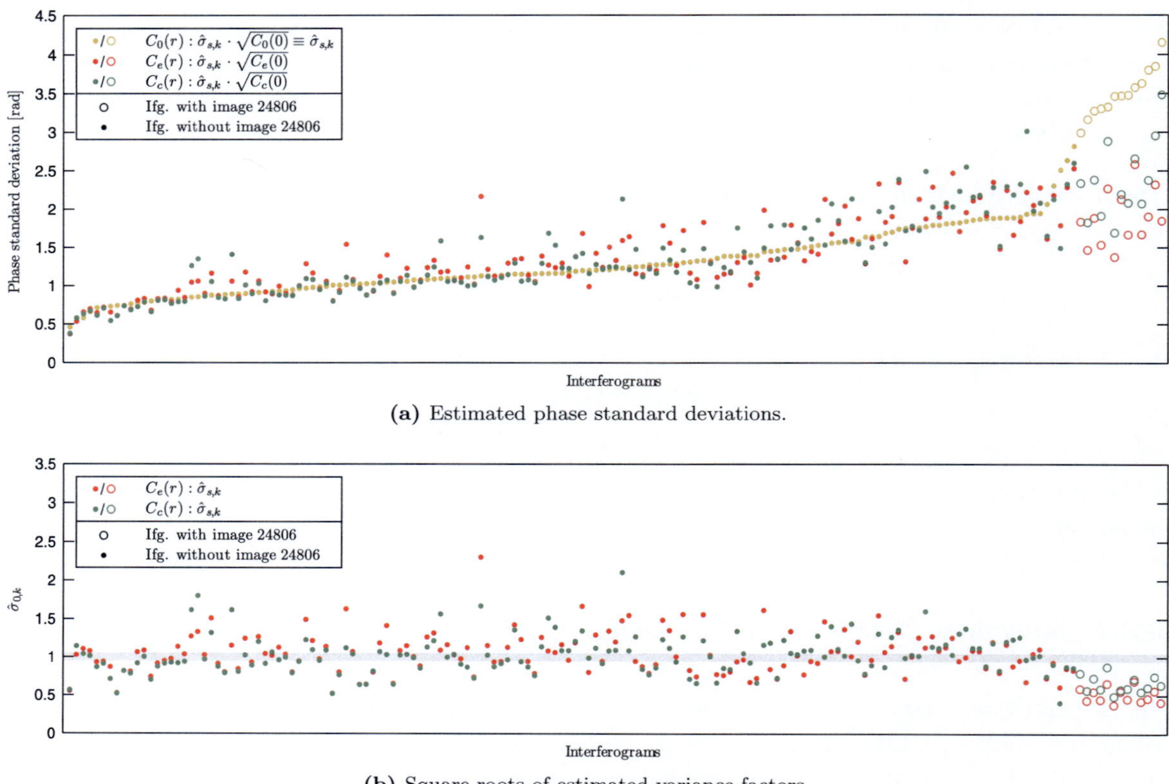

(a) Estimated phase standard deviations.

(b) Square roots of estimated variance factors.

Figure 6.3.: Overview of variance factor estimates for the 163 interferograms, ordered ascendingly by the estimates obtained with the simple covariance model $C_0(r)$ assuming uncorrelated observations. The estimates for one and the same interferogram are always arranged one below another in both subfigures. **(a)** $\hat{\sigma}_{s,k} \cdot \sqrt{C(r)}$ is an a posteriori estimate for the phase standard deviation. **(b)** $\hat{\sigma}_{s,k}$ is an indicator for the validity of the a priori assumed covariance model $C_e(r)$ or $C_c(r)$, respectively. The grey band marks the two-sided 99 % confidence region [0.95; 1.05] of the global model test. For the simple model $C_0(r)$, $\hat{\sigma}_{s,k}$ cannot serve for validation purposes, because the a priori variance $C_0(0) = 1$ is arbitrarily defined (see eq. (6.1)) and not estimated from the data; $C_0(r)$ is thus only considered in subfigure a. The variance components $\hat{\sigma}_{c,k}^2$ obtained in the closed approach are numerically almost identical ($|\hat{\sigma}_{s,k} - \hat{\sigma}_{c,k}|/\hat{\sigma}_{s,k} < 1.7 \cdot 10^{-3}$).

and the functional model (4.15) of orbital error mechanisms is well-defined by the acquisition geometry, the failed tests are considered an indicator for a deficient stochastic model.

The standard deviations of the estimated baseline errors range up to $\hat{\sigma}_{\delta\hat{B}_\parallel} = 0.95$ mm/s, $\hat{\sigma}_{\delta\hat{B}_\perp} = 14$ cm and $\hat{\sigma}_{\delta B_{\mathrm{fr}}} = 0.77$ fringes, respectively (see figure 6.4). These partly large values illustrate that the orbital error signal is generally superposed by a significant atmospheric contribution. The covariance models $C_e(r)$ and $C_c(r)$ are still crude approximations of the true stochastic behaviour of the interferometric phase, but they provide a closer approximation of reality than the simple model $C_0(r)$. Hence, the significantly higher standard deviations of baseline errors obtained with these models can be considered more realistic. Nevertheless, due to still unresolved model deficiencies, they serve rather as a rough indicator than a rigorous statistical quality measure.

Considering the least squares solution as an unbiased reference, the estimates obtained with the gridsearch method turn out to be unreliable in some cases. Figure 6.5a shows that there are large deviations of up to 5.7 fringes. These occur frequently in the presence of a nonlinear large-scale atmospheric signal and go along with the presence of more than one distinct local maximum in the search space (see figure 6.5b). Note that the least squares estimates do not need to coincide with a local maximum in the parameter search space, because both estimators are based on different objective functions ($\mathbf{v}_\varphi^T \mathbf{Q}_\varphi^{-1} \mathbf{v}_\varphi \to$ min versus $\left| \sum_{j=1}^{n_\varphi} e^{iv_{\varphi,j}} \right| \to$ max).

Table 6.1.: Results of sequential network adjustment. Individually estimated baseline errors $\delta\hat{B}$, adjusted orbit errors $\delta\hat{x}$, predicted corrections v_B and estimated biases $\hat{\nabla}B$ are characterised by their median and their maximum. The conversion to fringes follows eqs. (6.6) through (6.9). Red digits in the lower half of the table highlight numerical deviations with respect to the upper half.

				$\|\delta\dot{\hat{B}}_\|\|$ $[\frac{\mathrm{mm}}{\mathrm{s}}]$	$\|\delta\hat{B}_\perp\|$ [cm]	$\delta\hat{B}_\mathrm{fr}$ [fr.]	$\|\delta\dot{\hat{x}}_\|\|$ $[\frac{\mathrm{mm}}{\mathrm{s}}]$	$\|\delta\hat{x}_\perp\|$ [cm]	$\delta\hat{x}_\mathrm{fr}$ [fr.]	$\|v_{\dot{B}_\|}\|$ $[\frac{\mathrm{mm}}{\mathrm{s}}]$	$\|v_{B_\perp}\|$ [cm]	v_{B_fr} [fr.]	$\|\hat{\nabla}B_\mathrm{fr}\|$ [fr.]	rejected
Without data snooping	Least squares	$C_0(r)$	med	0.64	16.3	1.11	0.55	18.6	1.12	0.03	0.3	0.02		0
			max	3.43	94.8	4.78	2.68	65.2	4.08	0.30	2.0	0.22	0.26	
		$C_e(r)$	med	0.65	17.1	1.02	0.50	18.9	1.10	0.04	0.4	0.04		0
			max	3.12	91.6	4.43	2.31	65.0	3.83	0.47	3.5	0.33	0.35	
		$C_c(r)$	med	0.63	16.4	1.02	0.49	18.8	1.10	0.05	0.5	0.06		0
			max	3.11	91.6	4.43	2.39	65.6	3.93	0.50	4.5	0.39	0.50	
	Gridsearch		med	0.75	18.9	1.22	0.57	19.9	1.10	0.13	3.1	0.27		0
			max	5.50	120.2	7.12	3.87	72.1	4.07	3.78	88.1	5.61	6.68	
With data snooping	Least squares	$C_0(r)$	med	0.64	16.3	1.11	0.55	18.7	1.13	0.02	0.3	0.03		2
			max	3.43	94.8	4.78	2.67	65.2	4.08	0.16	2.0	0.15	0.17	
		$C_e(r)$	med	0.65	17.2	1.06	0.50	18.9	1.10	0.03	0.4	0.04		1
			max	3.12	91.6	4.43	2.31	65.0	3.83	0.47	3.5	0.32	0.35	
		$C_c(r)$	med	0.63	16.1	1.02	0.49	18.8	1.08	0.04	0.4	0.04		4
			max	3.11	91.6	4.43	2.31	65.0	3.85	0.56	3.3	0.40	0.43	
	Gridsearch		med	0.67	17.3	1.15	0.59	20.2	1.23	0.04	0.5	0.04		31
			max	3.71	81.9	4.94	2.92	71.6	4.45	0.19	2.7	0.16	0.25	

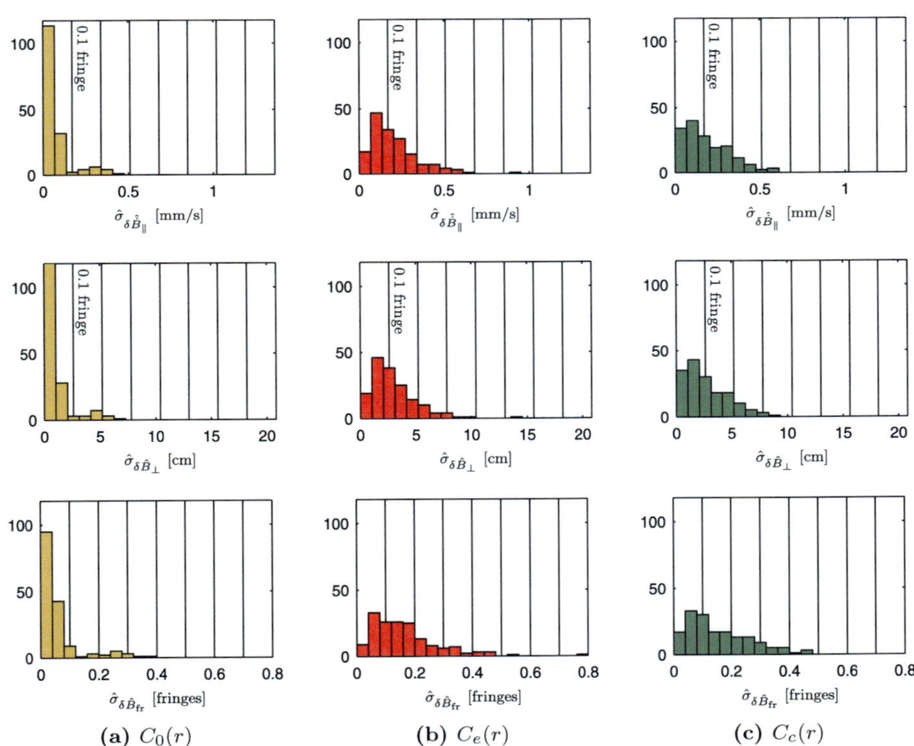

(a) $C_0(r)$ **(b)** $C_e(r)$ **(c)** $C_c(r)$

Figure 6.4.: Standard deviations of estimated baseline errors after individual, interferogram-wise estimation as described in section 4.3.1, using different covariance models. The conversion to fringes in the lower row follows eq. (6.6).

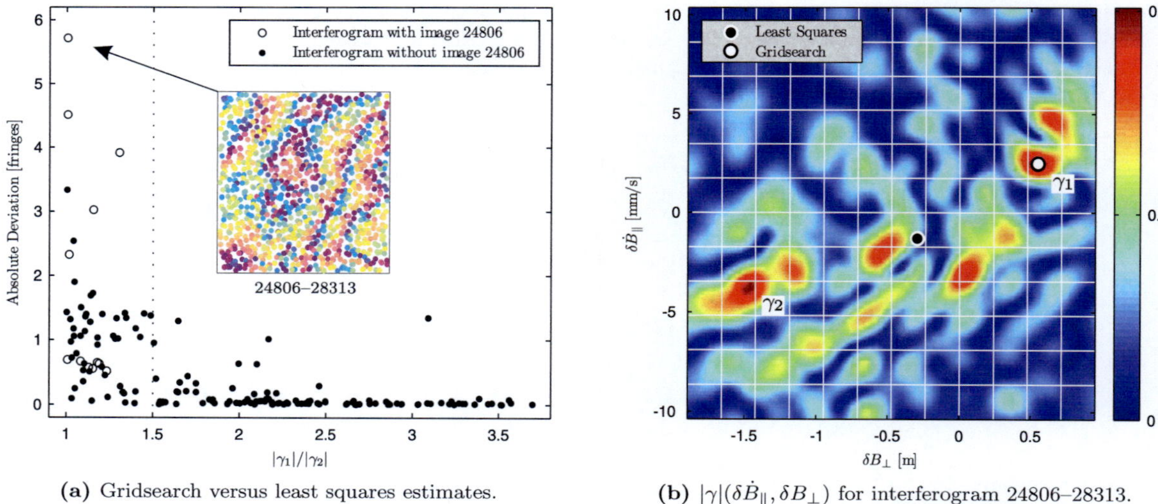

(a) Gridsearch versus least squares estimates.   **(b)** $|\gamma|(\delta\dot{B}_\parallel, \delta B_\perp)$ for interferogram 24806–28313.

Figure 6.5.: Comparison of gridsearch and least squares estimates ($C_0(r)$). **(a)** Absolute deviations have been computed with eq. (6.10) and are plotted as a function of the ratio γ_1/γ_2 between the highest and the second-highest local maximum in the search space $|\gamma|(\delta\dot{B}_\parallel, \delta B_\perp)$. The highest deviation has been observed in interferogram 24806–28313, the observations from which are visualised in the subframe. **(b)** Coherence measure $|\gamma|$ of interferogram 24806–28313 computed for incrementally varied baseline error values ($\delta\dot{B}_\parallel, \delta B_\perp$). The interval between the white grid lines corresponds to one fringe according to eq. (3.17). The gridsearch solution is defined by the highest value of $|\gamma|$, from which the least squares solution has a distance of 5.7 fringes – 2.2 in azimuth ($\delta\dot{B}_\parallel$) and 3.6 in range ($\delta B_\perp$). Note that due to the differing objective functions of the two estimators, the least squares solution does not necessarily coincide with a local maximum in the search space.

The ratio between the highest local maximum γ_1 and the second-highest local maximum γ_2 can be considered an indicator for the reliability of the gridsearch estimates. From figure 6.5a it can be seen that the probability of a biased estimate is high if $\gamma_1/\gamma_2 < 1.5$. In contrast to the least squares estimator, the gridsearch approach does not provide any statistically substantiated quality measure.

6.2.3. Adjusted Orbit Error Estimates

Adjusting baseline errors obtained with the least squares estimator yields absolute orbit errors of up to 2.7 mm/s in $\dot{x}_\parallel$ (1.6 fringes in azimuth) and 66 cm in $x_\perp$ (2.5 fringes in range; see table 6.1 and figure 6.6). Maximum corrections of $v_{\delta\dot{B}_\parallel} = 0.50$ mm/s (0.3 fringes) and $v_{\delta B_\perp} = 4.5$ cm (0.2 fringes) indicate an acceptable internal consistency of the network. This is different for the adjustment of the gridsearch estimates, where a maximum correction of $v_{B_{\mathrm{fr}}} = 5.6$ fringes illustrates that some individual baseline error estimates are strongly contradictory.

The choice of the covariance model for the least squares estimator determines the mutual weighting of individual contributions ($\delta\hat{B}_{\parallel,k}, \delta\hat{B}_{\perp,k}$) of interferograms but has only little effect on the adjusted orbit errors ($\delta\hat{x}_{\parallel,k}, \delta\hat{x}_{\perp,k}$). Their mutual deviations (6.11) computed for all interferograms have a median of 0.05 fringes and a maximum of 0.39 fringes (see table 6.2). They impressively illustrate the sensitivity with respect to the stochastic model assumptions. In light of this, the actual accuracy of estimates is probably not better than a few tenths of fringes.

As all interferograms have been processed with care and interferograms with dubious unwrapping results have preventively been eliminated from the network, the presence of large-scale unwrapping errors can basically be excluded. However, regardless the choice of the covariance function, there are interferograms that do not pass the outlier test $T_{s,k}$ (see eq. (5.32)) at a significance level of $\alpha = 0.1$ %. This can only be

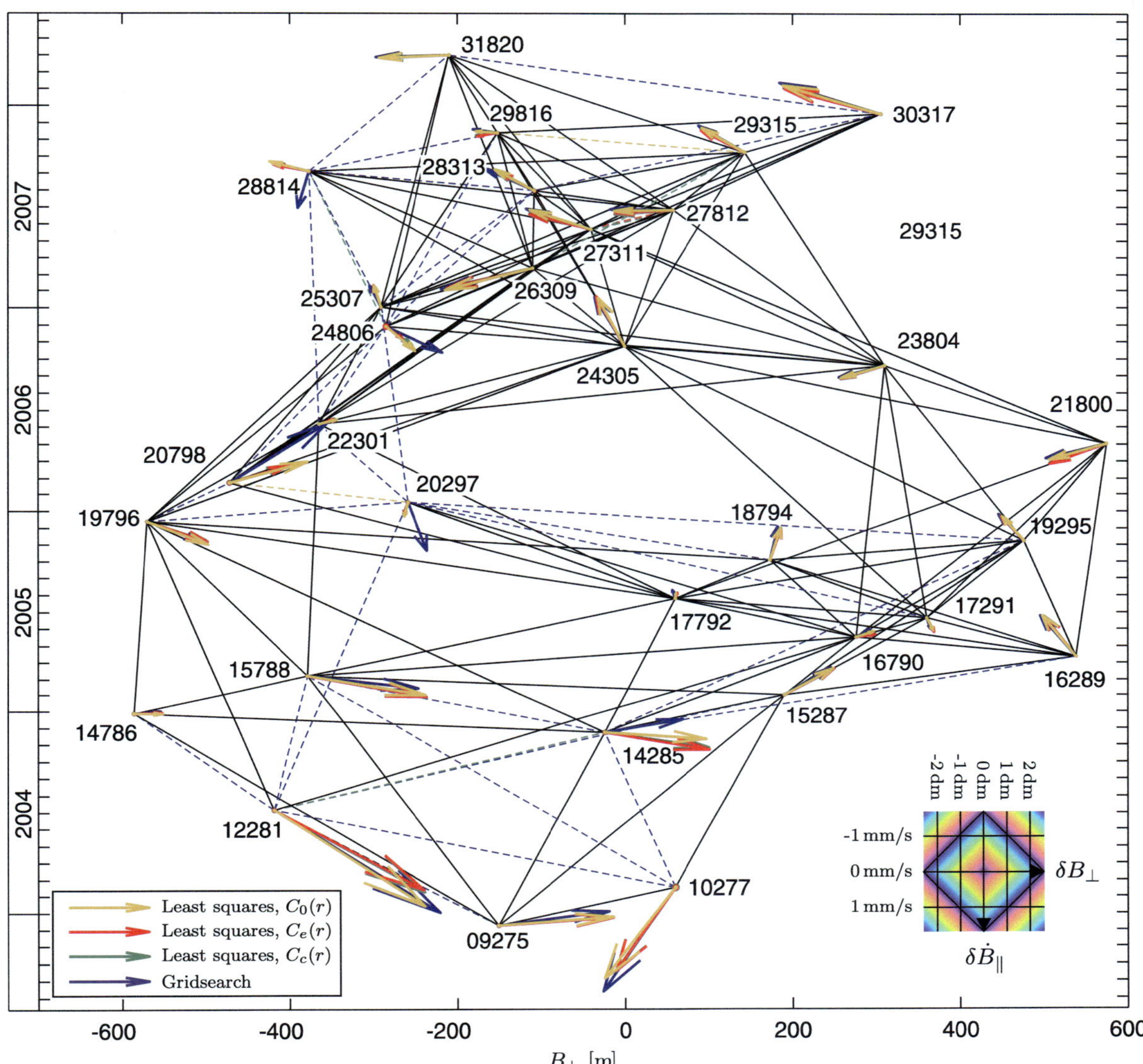

Figure 6.6.: Network of $n = 163$ interferometric combinations of $m = 31$ Envisat images. Vertices represent images (i. e., acquisitions with indicated orbit numbers), and edges stand for interferograms. The adjusted orbit errors after data snooping are represented by arrows visualising magnitude and orientation of the fringe gradient. The number of orbital fringes in the interferogram can be obtained from the legend in the lower right corner. The conversion follows eq. (3.17) and is based on the relation that one fringe in azimuth is equivalent to a baseline error of $\delta \dot{B}_\parallel = 1.7$ mm/s and one fringe in range corresponds to $\delta B_\perp = 26$ cm. These factors specifically apply to the analysed data set and the actual extent of the scene, which is why they deviate from the numbers in figure 3.5 and table A.3, respectively. The dashed edges represent interferograms that are identified and rejected as outliers during data snooping. With one exception (27311–27812), the four colour-coded approaches identify completely different interferograms as outliers.

Table 6.2.: Comparison of orbit error estimates obtained with different estimators and covariance models. For every pairing of approaches, both the maximum and the median absolute deviation in fringes between the estimated orbit error parameters are given; the absolute deviation is defined in eq. (6.11).

Median / Maximum		Without data snooping				With data snooping			
		Lsq. $C_0(r)$	Lsq. $C_e(r)$	Lsq. $C_c(r)$	Grid-search	Lsq. $C_0(r)$	Lsq. $C_e(r)$	Lsq. $C_c(r)$	Grid-search
Without data snooping	Lsq., $C_0(r)$		0.39	0.38	0.89	0.02	0.39	0.38	1.24
	Lsq., $C_e(r)$	0.05		0.24	1.14	0.40	0.02	0.06	1.29
	Lsq., $C_c(r)$	0.05	0.02		1.18	0.36	0.24	0.21	1.33
	Gridsearch	0.18	0.19	0.18		0.90	1.14	1.18	1.42
With data snooping	Lsq., $C_0(r)$	0.00	0.05	0.05	0.18		0.40	0.37	1.24
	Lsq., $C_e(r)$	0.05	0.00	0.02	0.19	0.05		0.06	1.29
	Lsq., $C_c(r)$	0.05	0.01	0.01	0.19	0.04	0.01		1.33
	Gridsearch	0.09	0.14	0.13	0.12	0.09	0.13	0.14	

Table 6.3.: Estimates for the global variance factor $\hat{\varsigma}_s^2$, given in terms of standard deviation. The two-sided 99 % confidence interval for $\hat{\varsigma}_s$ is $[0.89; 1.11]$.

Estimator	Least Squares			Gridsearch
Covariance model	$C_0(r)$	$C_e(r)$	$C_c(r)$	
$\hat{\varsigma}_s$ prior to data snooping	1.05	0.59	0.75	0.69
$\hat{\varsigma}_s$ after data snooping	1.01	0.59	0.66	0.20

explained by the remaining deficiencies of the stochastic model discussed in section 6.1. If data snooping is applied, only a small number of interferograms are rejected before all tests pass (see table 6.1 and figure 6.6). The consequential maximum change of the estimated orbit errors is only 0.02 fringes for the models $C_0(r)$ and $C_e(r)$ but 0.21 fringes for the model $C_c(r)$ (see table 6.2). Especially the latter result makes the data snooping technique seem questionable, considering that significant outliers are detected where no outliers are expected. Data snooping capabilities are analysed in more detail in section 6.3.

Applying data snooping to the inconsistent set of gridsearch estimates, 31 interferograms have to be rejected before all tests $T_{s,k}$ pass. With predicted corrections of 0.16 fringes and below, the consistency of the revised network is of a similar quality as the least squares solution (see table 6.1). Whereas both estimators yield individually consistent solutions that generally comply with one another, the estimates for some particular acquisitions deviate on the one-fringe level (see table 6.2). These acquisitions are either affected by significant nonlinear atmospheric trends or adjacent in the network graph to thereby affected acquisitions (see figure 6.6). Hence, the capability of the gridsearch estimator to reliably identify biased contributions of particular interferograms is limited to interferograms without significant nonlinear atmospheric trends.

In terms of estimated standard deviations, the adjusted orbit error parameters are significantly more precise than the individually estimated baseline error parameters (compare figure 6.7 with figure 6.4). This is an expectable consequence of the pseudo-redundancy in the overdetermined network. Considering the rather approximative than rigorous stochastic models, it is important to note that the standard deviations obtained from the covariance matrix $\hat{\varsigma}_s^2 \mathbf{Q}_{\hat{\mathbf{x}}_s \hat{\mathbf{x}}_s}$ are rather a rough indicator for the internal consistency than a precise measure of the absolute accuracy. This statement is additionally supported by the global variance factor estimates $\hat{\varsigma}_s^2$ (see table 6.3), for which $\mathrm{E}\{\hat{\varsigma}_s^2\} = 1$ and $\hat{\varsigma}_s^2 \sim F_{2(n-m+1),\infty}$ should apply. Validation by a global model test would require $\hat{\varsigma}_s \in [0.89; 1.11]$, but the test passes only for the simple covariance model $C_0(r)$. This is interesting, since $C_0(r)$ is the most rudimentary one of the evaluated covariance models, and one would expect it to perform worse than any alternative model.

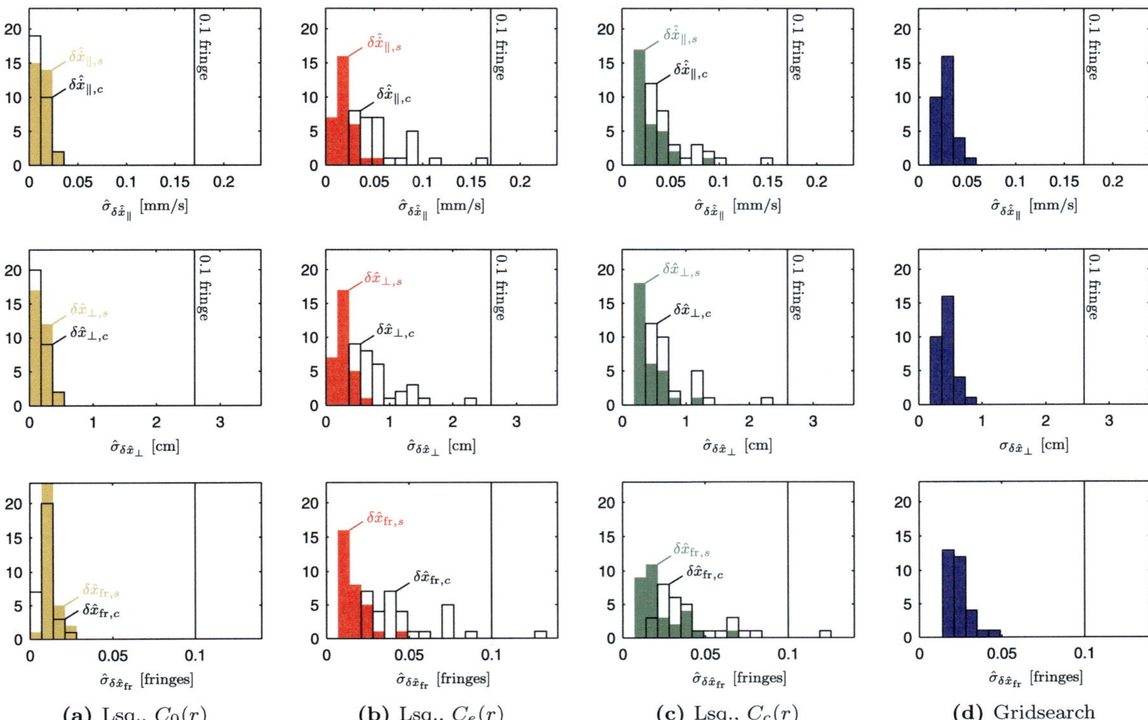

(a) Lsq., $C_0(r)$ (b) Lsq., $C_e(r)$ (c) Lsq., $C_c(r)$ (d) Gridsearch

Figure 6.7.: Standard deviations of estimated orbit error parameters. The conversion to fringes in the lower row follows eq. (6.7). **(a-c)** The standard deviations obtained with the least squares (lsq.) estimator are displayed for both the sequential ("s") and the closed ("c") approach prior to data snooping. Their ratio $\delta\hat{x}_s/\delta\hat{x}_c$ equals $\hat{\varsigma}_s$ (see eq. (5.11) and table 6.3) at a high level of numerical accuracy. **(d)** The standard deviations of the gridsearch estimates are displayed after data snooping. Otherwise, they would be extraordinarily large due to considerable misclosures in the network (see table 6.1).

6.2.4. Sequential versus closed Adjustment

The orbit error estimates obtained with sequential and closed adjustment can be considered numerically equivalent. The mutual deviation of estimated orbit errors is less than $1/1000$ fringe for all interferograms and for any of the three evaluated covariance models. The deviations can be explained by two minor simplifications of the sequential procedure: neglect of the convergence of viewing directions for master and slave (i. e., averaging of coefficients in eq. (4.7)) and biased variance factor estimates $\hat{\sigma}^2_{s,k}$ that are only based on intermediate, still inconsistent baseline error parameters (see figure 5.1a).

Also the corresponding cofactor matrices can be considered numerically identical for both approaches, i. e.,

$$\mathbf{Q}_{\hat{\mathbf{x}}_s\hat{\mathbf{x}}_s} \approx \mathbf{Q}_{\hat{\mathbf{x}}_c\hat{\mathbf{x}}_c} \,, \tag{6.12}$$

because all elements (i,j) satisfy:

$$\left| \frac{q_{x_{s,i}x_{s,j}} - q_{x_{c,i}x_{c,j}}}{\sqrt{q_{x_{s,i}x_{s,i}} \cdot q_{x_{s,j}x_{s,j}}}} \right| < 0.003 \,. \tag{6.13}$$

Nevertheless, there is one significant difference between the two approaches concerning the stochastic model: In the sequential approach, the cofactors are scaled by the global variance factor $\hat{\varsigma}^2_s$ to obtain the covariance matrix, whereas in the closed approach, there is no global variance factor (compare eq. (5.10) with eq. (5.21)). If introduced, its estimate would turn out to be one, because any dispersion of observations is already accommodated by the variance components $\hat{\sigma}^2_{c,k}$. Consequently, the standard deviations in the sequential approach equal the standard deviations in the closed approach multiplied by $\hat{\varsigma}_s$ (see figure 6.7).

In conclusion, the two approaches yield quasi-identical estimates with significantly different variance levels, both of them suffering from inconsistencies in the stochastic model. Hence, it cannot be decided, which one of them is more or less realistic.

6.3. Performance of Outlier Detection

As the detectability of outliers is considered the major advantage of the network approach, it is highly relevant to verify if the mechanisms developed in section 5.3 perform as expected. This question is initially approached from the theoretical point of view by validating if the dedicated test statistics $T_{s,k}$ and $T_{c,k}$, respectively, do indeed follow their associated statistical distributions. Subsequently, a more practical validation is concerned with the detection of simulated unwrapping errors in single interferograms.

6.3.1. Validation of Test Statistics

Two statistical tests have been proposed to detect unwrapping errors. On the interferogram level, the Student-distributed test statistic $T_{\varphi,i}$ (eq. (4.26)) indicates if an individual phase observation deviates significantly from the estimated orbital error signal. On the network level, a large-scale unwrapping error in an interferogram can be detected by means of the Fisher-distributed test statistics $T_{s,k}$ or $T_{c,k}$, respectively (eqs. (5.32) and (5.41)).

To assess the validity of the statistical distributions, Pearson's chi-square goodness-of-fit test (e. g., KREYSZIG, 1977, p. 229 et seq.) is applied to samples of $T_{\varphi,i}$ ($i = 1 \ldots n_{\varphi}$) and $T_{s,k}/T_{c,k}$ ($k = 1 \ldots n$). The respective samples are binned into $N = 15$ intervals, the size of which is adapted so that all bins contain an equal number of samples. Subsequently, the number h_i of tests in each interval is compared to the theoretical number $h_{i,0}$ of tests that is supposed to lie in that interval if the assumed statistical distribution applies. Finally, the goodness-of-fit test is based on the difference of these numbers:

$$T_{\chi^2} = \sum_{i=1}^{N} \frac{(h_i - h_{i,0})^2}{h_{i,0}} \sim \chi^2_{N-1} \,. \tag{6.14}$$

Test on the Interferogram Level

T_{χ^2} has been computed for all 163 interferograms (see figure 6.8) after iteratively rejecting a small number of not more than 30 outliers. This is necessary to guarantee that the samples are not contaminated by blunders, which is an indispensable prerequisite to validate the statistical distribution of the test. It turns out that even for a small level of significance $\alpha = 0.1$ %, the t-distribution can be validated for only 73 % of the interferograms if uncorrelated observations are assumed ($C_0(r)$). For the empirical models $C_e(r)$ and $C_c(r)$, the validation succeeds for only 23 % or 26 %, respectively. It is remarkable that the largest test statistics are obtained for interferograms formed with image 24806. This image is dominated by the most pronounced large-scale nonlinear signal (see figure 6.2d).

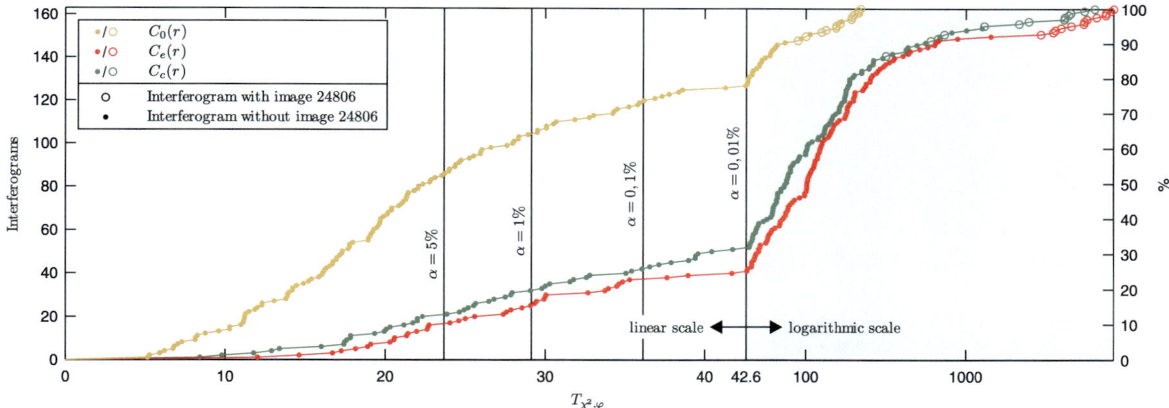

Figure 6.8.: Chi-square goodness-of-fit tests on the interferogram level. T_{χ^2} has been computed from the test statistics $T_{\varphi,i}$ of all $i = 1 \ldots n_\varphi$ pixels of an interferogram. The cumulative distribution of T_{χ^2} for all 163 interferograms is visualised by one graph per covariance model. The test fails for a large number of interferograms and any reasonable level of significance α.

It follows that the test is not capable to reliably identify isolated outliers in the phase observations and also reveals that neither of the three considered covariance models is qualified to adequately describe the stochastic behaviour of the observations. Whereas this latter conclusion is disturbing with regard to the overall model consistency, an underperforming detection of outliers in individual pixels can be tolerated as long as extraordinary large outliers are identified. Moderate biases in a small number of pixels do not distort the estimation significantly due to the large number of pixels it is based on.

Test on the Network Level

To validate the outlier tests on the network level, T_{χ^2} has been computed from the outlier tests of all 163 interferograms (see table 6.4). As a result, the validity of the Fisher distribution in eqs. (5.32) and (5.41) can be confirmed for the least squares estimator (see table 6.4) with no significant distinction between the rigorously derived test $T_{s,k}$ for the sequential approach and the pragmatically adopted test $T_{c,k}$ for the closed approach.

The Fisher distribution of $T_{s,k}$ turns out to be invalid for network adjustment of the gridsearch estimates. This outcome is no surprise, since the set of baseline error estimates is highly inconsistent (see table 6.1)

Table 6.4.: Chi-square goodness-of-fit tests on the network level. T_{χ^2} has been computed from the tests $T_{s,k}$ or $T_{c,k}$, respectively, of all $k = 1 \ldots n$ interferograms. Additionally, the maximum level of significance α is listed at which the Fisher distribution of the tests can be considered validated. The test is considered passed as long as $T_{\chi^2} < \chi^2_{14;0.95} = 23.7$ ($\alpha = 5\ \%$).

Estimator	Approach	Covariance model	Data snooping	T_{χ^2}	$\alpha\ [\%]$
Least Squares	Sequential	$C_0(r)$	no	21.5	9.0
		$C_e(r)$	no	22.0	7.9
		$C_c(r)$	no	15.6	33.8
	Closed	$C_0(r)$	no	20.5	11.4
		$C_e(r)$	no	23.3	5.6
		$C_c(r)$	no	15.2	36.1
Gridsearch	Sequential		no	735.6	0.0
			yes	54.2	0.0

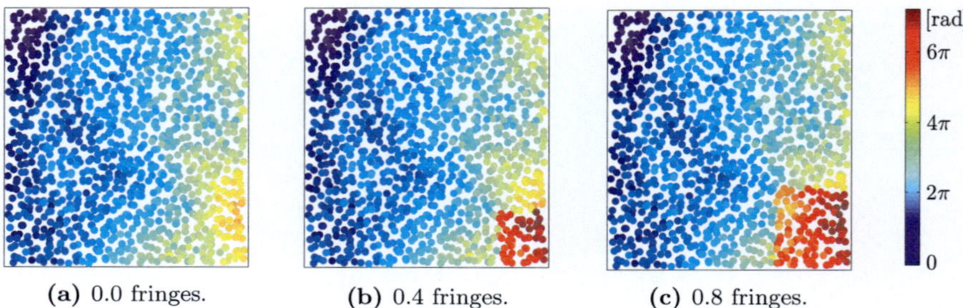

(a) 0.0 fringes.　　　　**(b)** 0.4 fringes.　　　　**(c)** 0.8 fringes.

Figure 6.9.: Simulation of unwrapping errors, exemplarily demonstrated on interferogram 14786–15788. All phase observations in a quadratically confined area in the lower right corner are shifted by 2π. The extent of the phase shifted area is measured by the fringe equivalent of the induced error signal.

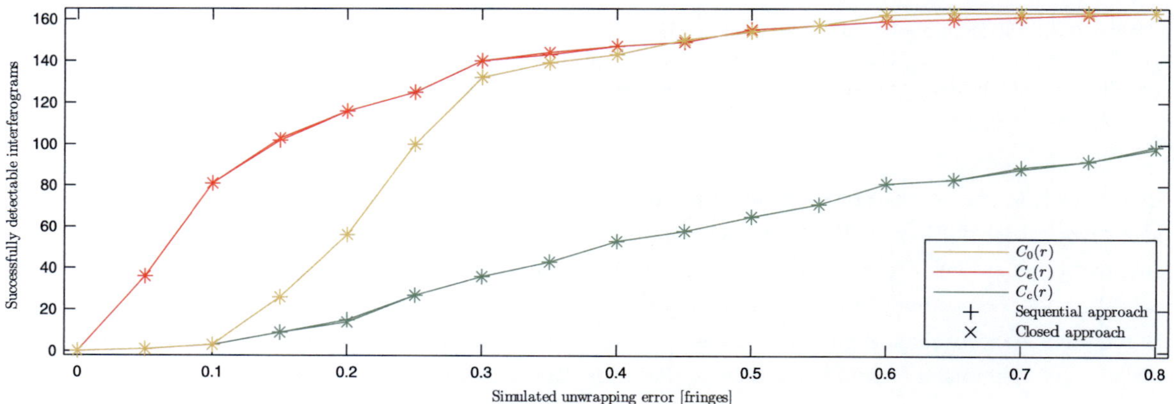

Figure 6.10.: Detectability of simulated unwrapping errors by the dedicated outlier tests $T_{s,k}$ and $T_{c,k}$, respectively. It is displayed in how many of the 163 interferograms a simulated error of a specified magnitude is detected while all other interferograms remain unchanged at the same time.

and thus probably contaminated by several outliers at a time. The goodness-of-fit test still fails after rejection of 31 potential outliers by data snooping. A likely explanation for this behaviour is the breakdown of the iterative data snooping procedure in presence of multiple outliers at a time masking each other. This illustrates the limitations of data snooping but does not allow the conclusion that the derived outlier test is generally useless for the adjustment of gridsearch estimates.

6.3.2. Detectability of Simulated Errors

To evaluate the sensitivity of the outlier test with respect to large-scale unwrapping errors, such errors have been simulated as demonstrated in figure 6.9. $n = 163$ case studies have been carried out, in each of which one of the 163 interferograms is contaminated by an unwrapping error with incrementally increased magnitude. The respective numbers of case studies, in which an unwrapping error of a particular magnitude is successfully detected are displayed in figure 6.10. An error simulated in interferogram k is considered detected if the test statistic for interferogram k both exceeds the critical Fisher fractile ($\alpha = 0.1$ %) and is larger than any other test statistic.

It turns out that the performance of outlier detection is very sensitive to the chosen covariance model, which basically determines the mutual weighting of interferograms in the network adjustment step. Comparing the models $C_e(r)$ and $C_c(r)$, it is remarkable that outliers are detected considerably faster with $C_e(r)$ although the covariance functions obtained for the two models are very similar (see figure 6.2).

Hence, no general conclusion can be drawn from this investigation to support the choice of a covariance model in general.

With the models $C_0(r)$ and $C_e(r)$, most unwrapping errors of 0.3 or more fringes are successfully detected. The finding that some errors above 0.5 fringes still remain undetected appears troubling but is relativised by the circumstance that the interferograms in question are associated with small weights in the network, mitigating their influence on the estimates. Nevertheless, the success rate obtained with the model $C_c(r)$ is definitely unacceptable.

Finally, it is worth noting that there are only small numerical differences between the test statistics $T_{s,k}$ for the sequential approach and the pragmatically adopted test statistics $T_{c,k}$ for the closed approach. Consequentially, their performance in outlier detection can be considered equivalent.

6.4. Conclusions and Outlook

6.4.1. Assessment of the proposed Methodology

The presented framework of orbit error estimation has been tested on a sample data set that was chosen due to its favourable premises for phase unwrapping. The test case cannot be considered optimal in all respects, mainly because the orbit error estimates are dominated by leakage from pronounced atmospheric trends. This circumstance compromises the demonstration of a best possible performance, but choosing a test area with calm atmospheric conditions would not have been representative for many practical applications, and engaging a more sophisticated approach for accommodating the atmospheric contribution would have exceeded the scope of this thesis. Whereas it is evident that the findings from this piece of research are not generalisable in all respects, they may allow some basic conclusions towards a generally applicable methodology.

A fundamental concern of the investigations was to find an appropriate stochastic model. Motivated by the assumption that correlated observations are a better approximation of reality, the use of empirical covariance functions has been evaluated. Their adjustment to individual interferograms turned out to be especially problematic for interferograms with distinct large-scale nonlinear atmospheric trends, which is probably the major reason for the failure of statistical validation; another one may be the neglect of algebraic correlation. In terms of results, the covariance model has only little effect on the estimates, but it significantly influences the corresponding dispersion measures, i. e., variances. However, these can neither be validated nor do they have any practical relevance except for outlier tests. There has been no evidence that the performance of outlier detection can be enhanced by considering correlations between observations either. Seeing the poor benefit in relation with the considerable effort of tailoring covariance functions, an orbit error estimation from uncorrelated observations appears to be an adequate compromise.

Sequential and closed approach can be considered equivalent, since the mathematical rigour of accounting for small differences in viewing directions of master and slave in the closed approach has no significant effect on the estimates. With the adaptions proposed in section 5.3.2, both approaches provide a practicable framework for outlier detection.

The gridsearch estimator provides a valuable methodology to infer orbit errors from the wrapped phase. However, it turned out to be more sensitive to large-scale nonlinear nuisance signals superimposed to the orbital contribution. Whereas the least squares estimator can accommodate these contributions in the stochastic model, they remain completely unconsidered by the gridsearch estimator, which does not provide a stochastic model at all. The consequential result was an inconsistent network of baseline error estimates. Nevertheless, the gridsearch method may be capable of producing more reliable results if the nonlinear atmospheric contribution at long wavelengths is either insignificant, mitigated beforehand or accounted for externally.

6.4.2. Potential Improvements

In spite of the basically promising results of the proposed approach, there are still some aspects that could be optimised by further research. One is the strategy of selecting phase observations that has to consider a trade-off between stochastical rigour and robustness. On the one hand, the observation quality is variable in space, and some parts of the interferogram might lack any reliable phase measurement. On the other hand, a quality-oriented selection or weighting of observations runs the risk that the estimates adjust to a local phase trend that is not representative for the whole interferogram. Whereas the approach of using unweighted and homogeneously distributed observations is an acceptable compromise, there may be a more sophisticated way to find an optimal compromise by means of robust estimation techniques (see section 4.5.1).

A major deficiency of the proposed least squares estimator is the still imperfect stochastic model. Especially the incidental violation of the stationarity assumption for the estimation of covariance functions and the neglect of algebraic correlations between interferograms are suspected to contribute significantly to the unsatisfactory performance. However, developing a more adequate model would be a complicated undertaking with uncertain benefit and thus not recommended in the first place. This is different for the gridsearch estimator, which does not involve a stochastic model at all. An adapted weighting scheme may have the potential to enable a more reliable outlier identification.

Regardless these conceivable enhancements, the greatest step forward in handling orbit errors could be made by applying the proposed methodology to a variety of data sets with different focusses of research. Thus, the performance in everyday applications can be evaluated, weaknesses identified, and strategies for further fine-tuning developed. Additional benefit may also be drawn from the joint consideration of orbit errors together with other signal components. Especially the deficient atmospheric modelling has potential for improvement by exploiting numerical weather models.

6.4.3. Embedding into the Processing Chain

There are several concepts on how to enhance the performance of InSAR processing by integration of orbit error estimation.

1. *A priori orbit correction.* Orbit errors are estimated without any consideration of deformation or atmosphere, and the predicted orbital signals are subtracted from the interferograms in a preprocessing step. This is the simplest approach and has been pursued within the scope of this thesis. Leakage of the long wavelength deformation component can be mitigated by temporal high-pass filtering of the orbit error estimates.

2. *Support of phase unwrapping.* Extraordinarily large orbit errors and the resulting spatial phase gradients between PS candidates can complicate or even impede phase unwrapping. As a remedy, approximate orbital error signals can be estimated, subtracted before and restored after the unwrapping step.

3. *Joint estimation with deformation and/or atmosphere.* The most elaborate methodology would be a joint estimation of all signal components, whereas the mathematical model for the orbital component could be based on the methodology presented here. As a joint estimation requires processing the data from all interferograms at a time, it involves a considerable computational load and is only practicable if the spatial dimension is reduced by hierarchical partitioning or pixel-wise estimation.

4. *Iteratively-alternating estimation with deformation and/or atmosphere.* Another option to reduce the complexity of the joint estimation (3.) is to alternatingly estimate the individual contributions and iterate towards converging parameters for deformation, orbit errors and eventually the atmospheric contribution. For the estimation of the orbital component, the methodology presented here can be used without adaption.

To facilitate the application of this methodology and related follow-up research, the estimators from chapter 4.3 have been integrated into the DORIS InSAR processor as an optional step (`http://doris.tudelft.nl`, version 4.04, step `ESTORBITS`, see also `M_MORBITS` and `S_MORBITS`).

6.4.4. General Applicability

Seeing the recent quality enhancements of orbit products, the need for orbit error estimation and correction is indeed becoming questionable. The trajectories for the latest SAR satellites have reached a level of accuracy at which the effect of residual orbit errors on interferograms is hardly significant. Moreover, modern InSAR processors are capable to filter out small orbital contributions by their spatio-temporal correlation properties, not requiring an explicit estimation (HOOPER, 2008; KETELAAR, 2009; FERRETTI ET AL., 2011). Nevertheless, there are a number of scenarios in which explicit orbit error correction will still be useful in the future.

1. *Radarsat.* With Radarsat-1 and Radarsat-2 there are two operational SAR satellites, the orbit accuracy of which is still far from meeting InSAR requirements. As strong Radarsat orbit errors may significantly complicate 3D phase unwrapping in PS processing, orbit error correction techniques are very relevant for processing Radarsat data. At least Radarsat-2 can be expected to operate for another couple of years, filling the gap in the series of C-band SAR missions between Envisat and Sentinel-1a.

2. *Historical missions.* Monitoring of long-term deformation processes always requires measurements spanning several years. At least during the next decade, there will still be considerable interest in acquisitions before 2010 or even from the 1990s to trace back the effects of anthropogenic activities or to maximise the temporal basis for the estimation of slowly creeping tectonic processes.

3. *Single interferograms or short time series.* If the data coverage is too poor to adequately support a temporal filtering, a simple orbit error correction may be reasonable if large-scale deformation signals are beyond the focus of research.

4. *Implicit temporal filtering.* Besides conventional filtering in the spatial or frequency domain, which is inherent to PS approaches, a distinction between deformation and orbit errors is alternatively feasible by a joint or iteratively-alternating estimation of both components with an adequate stochastic modelling. For the orbital component, the methodology proposed here can be used.

5. *Temporarily underperforming GPS.* The outstanding performance of GPS tracking cannot be guaranteed for any time in the future. An increase of ionospheric activity due to geomagnetic storms or

the general variations of the solar cycle may cause degradations in the quality of orbit products. It is also possible that GPS signals are artificially deteriorated at any time due to military considerations of the United States of America. This would entail a serious degradation of orbit determination performance as long as on-board receivers are not designed to complementarily treat signals of other GNSS. And though very unlikely, it cannot be excluded that the on-board GPS receiver of a SAR satellite fails, and orbits have to be determined with relatively imprecise backup systems.

6. *Near real-time applications.* The computation of precise ephemerides requires tracking data from both before and after an event as well as auxiliary data supporting various correction models. As acquisition, processing and gathering of these data takes some time, the more accurate orbit solutions are only available with a delay of some hours, days or even weeks. However, to support time-critical decisions in disaster management, it is not acceptable to postpone InSAR processing until high-quality orbit data are available. Hence, orbit error estimation techniques can be used to predict a precise trajectory based on interferometry with older images.

7. *Quality assurance.* Even with the expectable high quality orbits in the future, it will still be valuable to have a methodology in place to continuously check if the actual orbit accuracy meets the requirements.

6.4.5. Separability from other Signal Components

The interferometric signal can generally be decomposed into three contributions: deformation, atmospheric propagation delay and residual errors in the geometric phase, whereas the long wavelength component of the latter is dominated by orbit errors. For the estimation of either deformation, atmosphere or orbit errors, all three contributions have to be accommodated by either the functional or the stochastic model. Any imperfection in modelling may cause leakage from the interferometric signal of one component into the estimate of another component.

In order to assess the severity of leakage, the signal of interest has to be defined. In deformation analysis, deformation is the signal of interest. Estimating deformation parameters, leakage from the orbital contribution can be mitigated by estimating and subsequently subtracting the orbital error signal. Leakage can never be prevented completely, since it is infeasible to model the deformation signal both functionally and stochastically at an ultimate level of detail. Concrete strategies to distinguish the contributions of orbital errors and deformation have been outlined in section 4.2.2.

As the atmospheric contribution does not follow a characteristic pattern, it cannot be modelled functionally. A rigorous consideration in the stochastic model is only straightforward for the short scale component. The linear part of the large scale component is only separable from the orbital contribution by integrating complementary meteorological measurements and otherwise leaks into the orbit error estimates. Since deformation and not atmosphere is the signal of interest, this type of leakage is tolerable. However, a learning from section 6.2 is that large scale nonlinear atmospheric artefacts still can significantly bias the orbit error estimates and generate an inconsistent set of parameters, which is definitely not tolerable. Hence, consistent orbit error estimation can only go along with consistent atmospheric modelling, and the impact of leakage has to be assessed in context of the respective mathematical model.

7. Effect of the Reference Frame

Orbital state vectors are commonly expressed in a terrestrial reference frame with respect to which tectonic plates perform a relative, mostly horizontal motion in the order of some centimetres per year. In view of pixel sizes on the metre level it appears conclusive that the continuous large-scale plate displacements can safely be neglected for InSAR processing. However, this reasoning is only valid for the mapping of pixels to well-defined surface locations during the geocoding step. But state vectors are also required for the computation of the reference phase, which is computed from the range difference between master and slave acquisition. The two ranges are supposed to be measured from the respective sensor positions to a *common* target on the surface (see figure 2.2), and it would not be sound to presume lower accuracy requirements for the target identity than for the sensor locations. A relative orbit accuracy on the centimetre level consequently implies a *relative* target accuracy in the same order. The latter is not provided though if the target performs a typical tectonic displacement between two acquisitions that are separated in time by a couple of years.

Another explanatory approach for the sensitivity of interferometry to tectonic motion is based on a closer examination of the acquisition geometry. A premature conclusion might be that no deformation signal can be expected from a rigid tectonic plate performing an almost translational motion. A homogeneous displacement would only induce a global phase shift that is irrelevant for the relative interferogram. However, due to the diverging lines of sight (LOS) at different ranges, the resulting phase shift is not necessarily homogeneous. Whereas a global surface displacement towards or away from the sensor is indeed almost invisible in the interferogram, a displacement perpendicular to the LOS induces a phase trend in range (see figure 7.1a).

If this kind of phase trend is considered a signal or an artefact depends on the point of view or the geodetic datum, respectively. Seen from the reference frame of the orbits, the interferogram reflects precisely the observed displacement. However, the definition of this frame is in a way arbitrary, and it is usually intended to measure only displacements relative to an assumedly motionless tectonic plate. This measurement can be obtained by relocating the point of view onto the plate (see figure 7.1b). Observed from there, the satellite orbit or its reference frame, respectively, performs a relative motion with respect to the surface. Compensating for this motion by a datum transformation of the orbital state vectors yields an unbiased interferometric measurement of displacements relative to the observed tectonic plate.

Neglecting this compensation in InSAR processing is comparable to making an error in the interferometric baseline, the size of which is increasing with the temporal baseline. It has been shown in section 3.4.1 that an error in $B_\perp$ of 26 cm induces one fringe in range in a full ERS or Envisat interferogram. In a reverse conclusion, an error signal of one fringe can likewise be expected if the tectonic plate motion perpendicular to the LOS accumulates to the same amount. As this may indeed happen for temporal baselines of only a few years, the compensation for tectonic motion can be considered definitely relevant for InSAR processing.

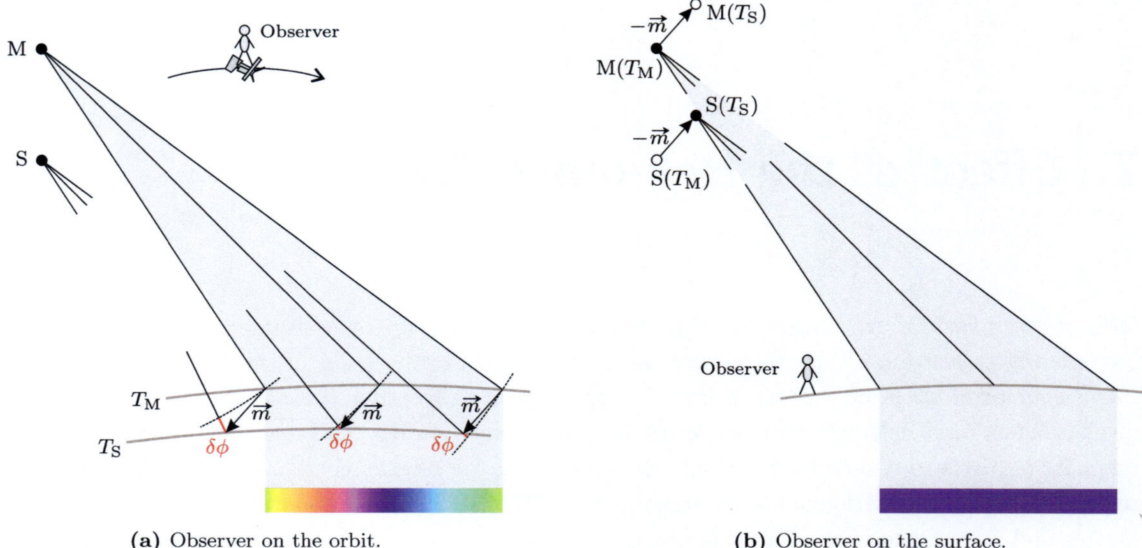

(a) Observer on the orbit.   **(b)** Observer on the surface.

Figure 7.1.: Interferometric signal of tectonic plate motion observed from different points of view. **(a)** Satellite orbits are given in a terrestrial reference frame, with respect to which the illuminated tectonic plate makes a relative movement $\vec{m}$ between the two acquisitions at T_M and T_S. Due to diverging LOS for different ranges, a phase trend in range is induced. **(b)** Observed from the perspective of a motionless tectonic plate, the orbital reference frame performs an apparent movement $-\vec{m}$ between the acquisition times T_M and T_S of master and slave, respectively.

The effect of relative tectonic plate motion with respect to the orbit reference frame on SAR interferograms has been introduced as the *reference frame effect* (BÄHR ET AL., 2012). In this chapter, a qualitative and quantitative evaluation is provided, complemented by a discussion of conceivable correction approaches.

7.1. Characterisation

Before describing and evaluating the reference frame effect in detail, a brief introduction to the concept of terrestrial reference frames is provided.

7.1.1. Terrestrial Reference Frames

Terrestrial reference frames (TRF) are realisations of earth-fixed coordinate systems and defined by three-dimensional positions $\vec{x}$ of a set of terrestrial stations. Modern TRF definitions relate these positions to a specified reference epoch T_ref and additionally comprise linear velocity components $\vec{v} = \vec{x}' = \partial\vec{x}/\partial T$ (recall eq. (2.3)) to account for continuous tectonic motion:

$$\vec{x}(T) = \vec{x}(T_\mathrm{ref}) + (T - T_\mathrm{ref})\vec{v} \,. \tag{7.1}$$

The stations generally coincide with reference points of geodetic instruments like GNSS antennas, radio telescopes, laser devices and radar beacons. The geodetic datum of a TRF has up to fourteen degrees of freedom: three translations (T_x, T_y, T_z), one scale factor D, three rotations (R_x, R_y, R_z) and the rates of change of these seven parameters $(T_x', T_y', T_z', R_x', R_y', R_z', D')$. A set of orbital state vectors can be expressed in any TRF, whereas a datum transformation from one TRF to another can be implemented by a similarity transformation with up to 14 parameters (BÄHR ET AL., 2007, ch. 4). However, it is important to note that a transformation between TRF is not unique, since the number of identical point coordinates generally exceeds the number of degrees of freedom.

As orbit determination is based on space geodetic techniques, the involved ground observing stations provide direct access to a TRF. Hence, it is straightforward to express an orbit solution in any TRF to the definition of which all observing stations contribute. The underlying TRF of an orbit solution forms part of the many assumptions and models that constitute the processing strategy of orbit determination. Thus, Envisat orbits from the French *Centre National d'Etudes Spatiales* (CNES) or the *European Space Operations Centre* (ESOC) are expressed in the *International Terrestrial Reference Frame* (ITRF; M. Otten, ESOC, pers. comm., 2010).

The ITRF is the most elaborate global reference frame and has evolved into a widespread standard for scientific applications. Not all orbit solutions are expressed in the ITRF, but many alternative TRF can be considered equivalent at some level of accuracy. TerraSAR-X annotated orbits for instance are expressed in the TRF "WGS84-G1150" (SCHUBERT ET AL., 2012), which is coincident with the ITRF at the 10 cm level (IGN, 2007). The latter statement applying to station positions, station velocities of WGS84 realisations and the ITRF may be assumed to be coherent at a high level of accuracy. Consequently, all further considerations and predictions within this chapter can be based on the ITRF without significant loss of generality.

The ITRF is a realisation of the conventionally defined *International Terrestrial Reference System* (ITRS; PETIT AND LUZUM, 2010). ITRF station positions and velocities are based on observations from space geodetic techniques like *Very Long Baseline Interferometry* (VLBI), SLR, GNSS and DORIS (see section 3.3.1). Starting with ITRF88, twelve releases have been published to date, continuously adding new observations and refining estimation strategies. The most recent ones are ITRF2000 (ALTAMIMI ET AL., 2002), ITRF2005 (ALTAMIMI ET AL., 2007) and ITRF2008 (ALTAMIMI ET AL., 2011). The oldest observations date from about 1980, and the number of the respective release approximately specifies the year of the latest observations included. ITRF2008 comprises positions and velocities of 935 stations. For some sites, multiple solutions (7.1) have been estimated for time spans separated by discontinuities like tectonic events or antenna changes.

The ITRF datum is based on the following conventions: The three translations and their rates are defined by the centre of mass of the earth sensed by SLR, and the scale and its rate are fit to the metre convention via SLR and VLBI measurements (ALTAMIMI ET AL., 2011). The orientation of the frame is basically arbitrary, aligning the three orientation parameters and their rates recursively to preceding ITRF realisations (ALTAMIMI ET AL., 2002, 2007, 2011). For ITRF2000, however, the orientation rates have been explicitly aligned to the geophysical plate kinematic model NNR-NUVEL-1A (ALTAMIMI ET AL., 2002; DEMETS ET AL., 1994). Consequently, this alignment applies recursively to the subsequent releases ITRF2005 and ITRF2008.

7.1.2. Prediction

Figure 7.2 illustrates how neglecting relative tectonic motion can bias the computed reference phase. For this purpose, a target P situated in the rigid interior of a tectonic plate is considered. The sensor positions M for the master acquisition at time T_M and S for the slave acquisition at time T_S are given with respect to the ITRF. If P is assumed to maintain its position from T_M in the coordinate frame of the ITRF (figure 7.2a), the simulated line of sight for the slave acquisition (dashed in the figure) does

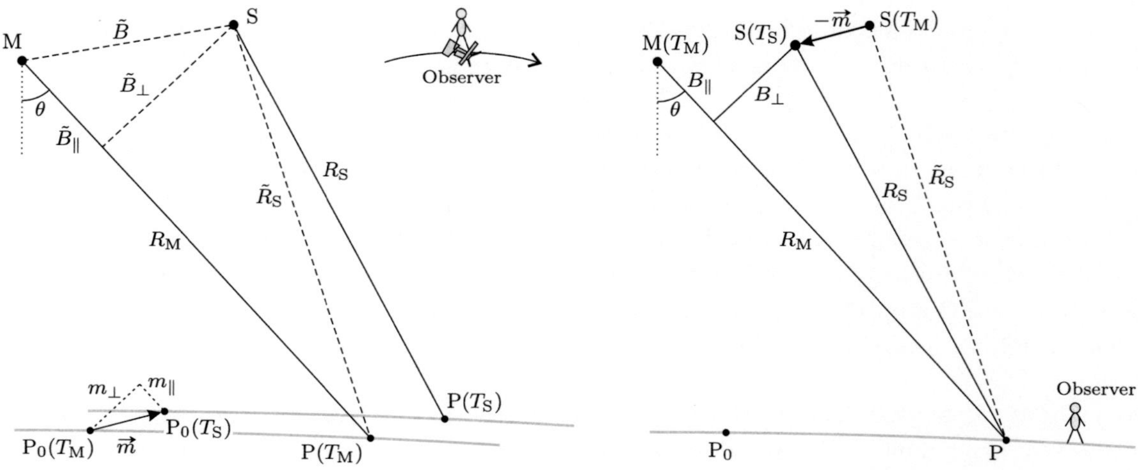

(a) Acquisition geometry in the ITRF datum.

(b) Acquisition geometry in the reference system of a motionless tectonic plate.

Figure 7.2.: Geometry of a master acquisition from orbit M at time T_M and a slave acquisition from orbit S at time T_S. **(a)** The sensor positions M and S are given in ITRF coordinates, whereas the tectonic plate with the acquired region performs a relative motion with respect to the ITRF. Assuming a non-deforming plate, this motion can be described by the displacement vector $\vec{m} = (T_S - T_M)\vec{v}$ of a nearby ITRF station P_0. If the motion is neglected for the computation of the reference phase, the biased range $\tilde{R}_S$ is used in eq. (3.13), implying a biased perpendicular baseline $\tilde{B}_\perp$. **(b)** Observing from a viewpoint on the tectonic plate, ITRF coordinates perform a shift, which has to be applied to the orbit data to yield an unbiased reference phase.

not reflect the actual range measurement R_S. Using $\tilde{R}_S$ for the computation of the reference phase instead of R_S results in a biased baseline $\tilde{B}$.

To compute an unbiased reference phase, the orbital state vectors have to undergo a datum transformation from the ITRF datum to a frame in which the tectonic plate under consideration is static. The simplest realisation for such a transformation assumes that the plate motion is only translational and can be adequately described by the velocity vector $\vec{v}$ of a nearby ITRF station P_0:

$$\vec{x}_{\text{plate}}(T) = \vec{x}_{\text{ITRF}} - (T - T_0)\vec{v} \,. \tag{7.2}$$

T_0 is the epoch at which the two frames coincide. Its choice is almost arbitrary, since a homogeneous shift of the state vectors of both acquisitions does not change the baseline. In figure 7.2b, $T_0 := T_M$ has been chosen, which means no change to the master orbit M and a shift of $-\vec{m} = -(T_S - T_M)\vec{v}$ to the slave orbit S. The reference phase computed from the transformed orbit positions $M(T_M)$ and $S(T_S)$ is unbiased, implying the actual perpendicular baseline $B_\perp$. Thus, the error $\delta B_\perp$ in the perpendicular baseline due to neglecting the reference frame effect can be predicted from the component of $\vec{v}$ perpendicular to the line of sight:

$$\delta B_\perp = \tilde{B}_\perp - B_\perp = (T_S - T_M)\,v_\perp = B_T\,v_\perp \,. \tag{7.3}$$

The baseline error component $\delta B_\parallel$ in ranging direction can be ignored, since it does not affect the interferometric phase in a significant way (see figure 3.5a). The maximum phase bias in range ("rg") can be predicted according to eq. (3.15):

$$\Delta_{\text{rg}}\delta\phi = -\Delta_{\text{rg}}\delta\phi_{\text{ref}} = \frac{4\pi}{\lambda}\,B_T\,v_\perp\,\Delta\theta \,, \tag{7.4}$$

where $\Delta\theta$ is the look angle difference between near range and far range. For instance, $\Delta_{\mathrm{rg}}\delta\phi_{\mathrm{ref}} = 2\pi$ would imply an almost linear error signal of one fringe in range (see figure 3.5b). Translating $\Delta_{\mathrm{rg}}\delta\phi_{\mathrm{ref}}$ into a bias of the estimated ground displacement rate D' in the line of sight yields with eqs. (2.3) and (2.9):

$$\Delta_{\mathrm{rg}}\delta D' = -v_\perp \, \Delta\theta \ . \tag{7.5}$$

Considering the sign convention from section 2.2.2, a positive $\Delta_{\mathrm{rg}}\delta D'$ implies a tilt of the ground towards the sensor and vice versa. However, this apparent interpretation does not reflect the actual cause of the signal, which is a translational motion of the tectonic plate perpendicular to the line of sight. This translation is misinterpreted as a tilt if the reference frame effect is not corrected for.

7.1.3. Global Evaluation

To get a global picture of the reference frame effect, the associated baseline error rate $\delta B'_\perp = v_\perp$ has been predicted for 840 of the 935 ITRF2008 stations that qualify by an accurate velocity estimate, preferably representative for a long timespan. 95 stations have been disregarded, because either their observations span less than a year or the standard deviation of their 3D velocity $|\vec{v}|$ exceeds 1 mm/a. For stations with multiple solutions, referring to different time spans separated by tectonic events or antenna changes, always the solution with the longest observation period has been selected.

During the 35 day repeat cycle of Envisat, each station is covered by several swaths. The evaluation of the reference frame effect has been performed for one ascending and one descending track in IS2 mode, for each of which the station is closest to the middle of the swath. For both, the error rate $\delta B'_\perp = v_\perp$ in the perpendicular baseline has been predicted from the ITRF velocity vector $\vec{v}$, where the decomposition into $v_\parallel$ and $v_\perp$ is defined by the line of sight to the middle of the swath.

Figure 7.3 and table 7.1 give an overview of the predicted baseline error rates. For most plates the effect behaves largely homogeneous, in some cases undergoing smooth variations due to rotational plate motion, for instance on the Australian plate. Only in deforming zones like the Andes or Japan, the rates follow a distinctly different pattern or appear even arbitrary. Hence, except for some regions, a prediction of the reference frame effect is expected to perform well with the velocity vector of the closest ITRF station.

The largest baseline error rates of 6 cm/a are predicted for descending tracks on Hawaii. If the effect is not corrected for, this would cause a phase ramp in range equivalent to a relative difference of $\Delta_{\mathrm{rg}}\delta D' = 7$ mm/a in the displacement rates observed at near range and far range. For a temporal baseline of four years, the error signal would already amount to one fringe. Other regions where the effect is very large are Baja California (Mexico), southern California (USA), the Indian plate and western Australia. The predicted baseline error rate is also considerable for some smaller Islands in the Pacific Ocean, but the associated error signal would be less pronounced due to the limited extension of land masses.

All predictions from this section also apply to the ERS satellites, which followed the same orbit as Envisat. For other sensors, similar results are expected. The most determining factor is the orientation of the perpendicular component $B_\perp$ of the interferometric baseline, which is defined here by orthogonality with respect to both the satellite trajectory and the line of sight. As SAR satellites commonly have a sun-synchronous orbit with an inclination around 98°, all of them have similar local headings. More variable is the respective line of sight, since the look angle typically varies between 15° and 60°. As plate motions are dominated by their horizontal component, the reference frame effect is expected to be larger for steep looking beam modes, for which the orientation of $B_\perp$ is rather horizontal than vertical. Finally, the bias of deformation estimates due to baseline errors increases with the swath width, which is owed to the ramp-like characteristic of the error signal. This is an important conclusion in view of the planned mission Sentinel-1, where the *Interferometric Wide Swath Mode* is designed with a swath width of 250 km.

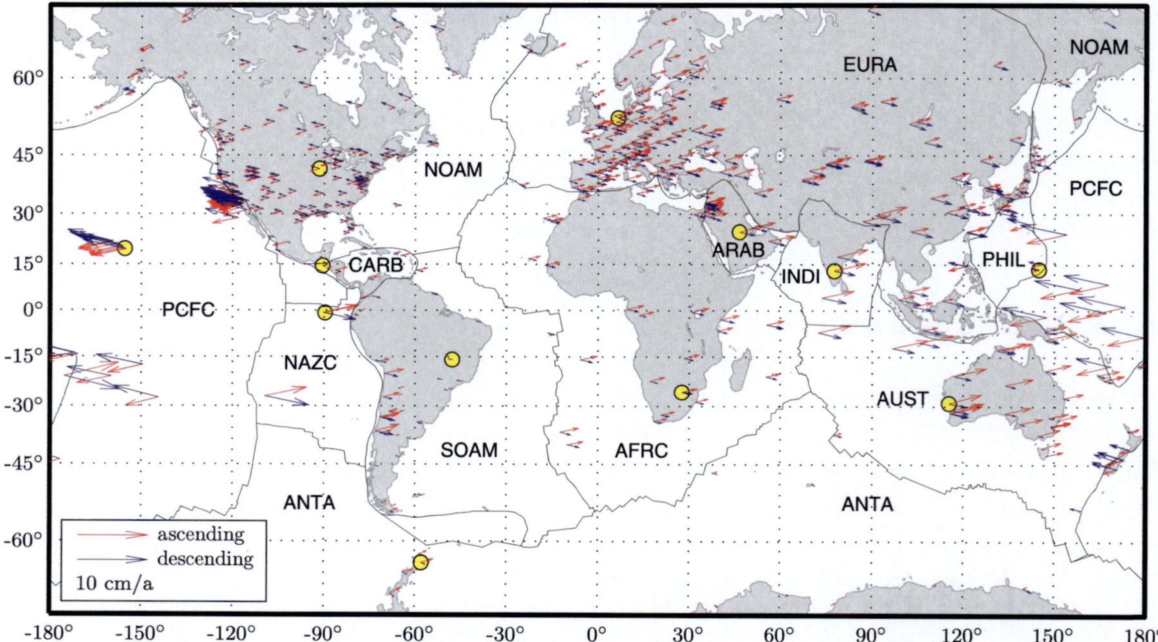

Figure 7.3.: Prediction of baseline error rates $\delta B'_\perp = v_\perp$ for Envisat IS2 interferograms from ITRF velocities, evaluated at 840 ITRF2008 stations. The arrows are aligned to the horizontal projection of the respective radar LOS. Numerical values for the twelve encircled stations can be found in table 7.1.

Table 7.1.: Prediction of the reference frame effect observable in Envisat IS2 interferograms for one sample ITRF station per tectonic plate. Figure 7.3 provides information on their representativeness. Each station has a unique "DOMES" identifier and may have multiple solutions for different time spans separated by tectonic events or antenna changes. $v_\perp$ is the baseline error rate, predicted from ITRF2008 and NNR-NUVEL-1A velocities, respectively. $\Delta_{\mathrm{rg}}\delta D'$ is the bias of the relative LOS displacement rate between near range and far range. The last column contains the temporal baseline $B_{T,2\pi} = |\lambda/(2\,\Delta_{\mathrm{rg}}\delta D')|$ that is required to induce an error signal of one fringe in range. All numbers are provided for both ascending and descending orbits.

Plate	Site name	DOMES no.	Solution no.	$v_\perp$ ITRF2008 [mm/a]		$v_\perp$ NUVEL-1A [mm/a]		$\Delta_{\mathrm{rg}}\delta D'$ ITRF2008 [mm/a]		$B_{T,2\pi}$ ITRF2008 [a/fringe]	
				asc.	desc.	asc.	desc.	asc.	desc.	asc.	desc.
AFRC	Hartebeesthoek (ZA)	30302S001	1 (1986-2008)	20	-12	23	-14	-2.2	1.3	13	22
ARAB	Riyadh (SA)	20101S001	1 (2001-2008)	36	-23	33	-19	-3.9	2.4	7	12
ANTA	O'Higgins (−)	66008S001	1 (1993-2008)	17	-6	18	-9	-1.8	0.7	15	41
AUST	Yarragadee (AU)	50107M001	1 (1982-2008)	48	-25	48	-25	-5.2	2.7	5	10
CARB	Guatemala City (GT)	40901S001	1 (2000-2009)	6	-4	4	-4	-0.6	0.4	46	63
EURA	Westerbork (NL)	13506M005	1 (1997-2009)	21	-10	21	-12	-2.2	1.1	13	25
INDI	Bangalore (IN)	22306M002	1 (1996-2005)	46	-30	46	-28	-5.0	3.2	6	9
NAZC	San Cristobal (EC)	42004S001	1 (1993-2005)	49	-45	59	-55	-5.3	4.8	5	6
NOAM	North Liberty (US)	40465S001	1 (1992-2008)	-15	13	-15	14	1.6	-1.4	18	20
PCFC	Mauna Kea (US)	40477S001	1 (1993-2008)	-50	61	-46	58	5.4	-6.6	5	4
PHIL	Guam (US)	50501S001	1 (1993-2002)	-7	10	-42	42	0.8	-1.1	35	26
SOAM	Brasilia (BR)	41606M001	1 (1996-2007)	-1	6	-2	6	0.1	-0.6	352	45

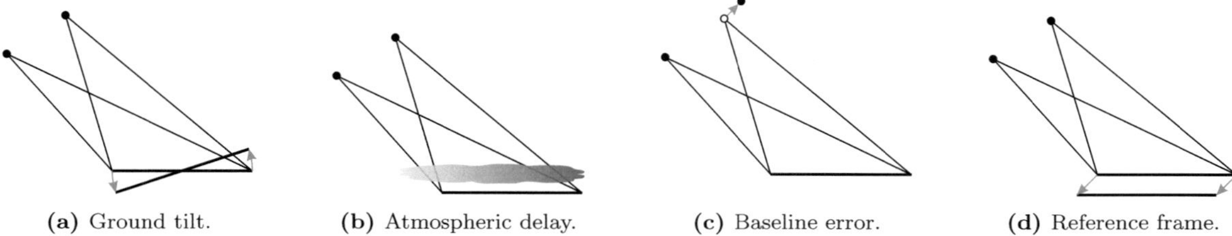

(a) Ground tilt.   (b) Atmospheric delay.   (c) Baseline error.   (d) Reference frame.

Figure 7.4.: Possible causes of a linear phase ramp in range. **(a)** Ground tilt or relative displacement in LOS direction, respectively. **(b)** Gradient atmospheric propagation delay. **(c)** Relative orbit error or baseline error, respectively. **(d)** Homogeneous ground displacement perpendicular to the LOS due to relative tectonic motion with respect to the orbit reference frame.

7.2. Consideration and Correction

An observed linear phase ramp in range can have different causes. Interpreting the signal as deformation, it would indicate a tilt of the ground with a relative displacement of $\lambda/2$ per fringe. But it can also be caused by a baseline error, a gradient atmospheric delay or lateral tectonic motion with respect to the orbit reference frame (see figure 7.4). The phase sensitivity to baseline errors and lateral displacement is equivalent and amounts to $\lambda/(2\,\Delta\theta)$ per fringe according to eq. (3.17). Hence, current sensors with relatively narrow fields of view ($\Delta\theta = 2°\ldots6.3° = 0.03\ldots0.11$ rad) are considerably more sensitive to ground tilts than to lateral tectonic displacement, and it is reasonable to assume rather a tilt than a lateral displacement when observing a ramp in range. For Sentinel-1 IWS with $\Delta\theta = 14.4° = 0.25$ rad this disproportion will be less pronounced.

Modern InSAR processing approaches distinguish between signal components by their spatio-temporal correlation properties. Atmospheric and orbital contribution can be separated from the deformation signal, because they are uncorrelated in time. Unfortunately, both large-scale deformation signals (i. e., ground tilts) and plate tectonic motion are correlated in both time and space, making a distinction by correlation properties infeasible. Hence, the reference frame effect would be fully absorbed by deformation estimates if it is not accounted for. But in contrast to deformation signals, the reference frame effect is well predictable and can thus be corrected for.

If it is appropriate to apply a correction or not, depends on the reference frame in which the measured ground displacements are supposed to be expressed. If this was the reference frame of the orbit data, no correction would be required. However, the InSAR technique is only capable to sense relative displacements with respect to a dedicated reference point on the surface. If this point was introduced in the orbit reference frame, it would undergo a continuous displacement due to the tectonic motion of the plate it is situated on. Thus, all measured displacement rates would be subject to a constant offset, complicating further interpretation. It is rather convenient to assume a motionless reference point and thus relate all measured displacement rates to this point. This can be achieved in processing by pre-transforming all involved orbital state vectors to a geodetic datum in which the assumedly rigid tectonic plate of the reference point is static.

Not all of the 14 possible degrees of freedom (T_x, T_y, T_z, D, R_x, R_y, R_z, T'_x, T'_y, T'_z, D', R'_x, R'_y and R'_z; see section 7.1.1) are relevant for a dedicated datum transformation. Since InSAR is a relative technique, any effect of the first seven parameters would cancel out in processing making their choice arbitrary. The scale factor is defined by the metre convention and thus invariant. Consequently, the interferogram is only sensitive to the six remaining parameters, i. e., three translation rates and three rotation rates. In

Table 7.2.: Overview of the three proposed types of datum transformation to correct for the reference effect with up to six degrees of freedom (DOF).

Approach		Translation	Euler Rotation	General Transformation
Parameters		T'_x, T'_y, T'_z	Φ, Λ, ω	$T'_x, T'_y, T'_z, R'_x, R'_y, R'_z$
DOF	considered	3 translation rates	2 Euler pole coordinates 1 angular velocity	3 translation rates 3 rotation rates
	neglected	3 rotation rates	1 subsidence/uplift 2 tilt rates	—
Approxima- tion quality	local	good	mostly acceptable	optimal
	global	degrading	acceptable	optimal
Data Source		ITRF station velocities	plate kinematic model	(GNSS) velocity field

the following subsections, three different correction approaches are proposed and discussed that dispose of these six degrees of freedom or a subset thereof, respectively (see table 7.2).

7.2.1. Translation

The most intuitive correction approach that accounts for the three translation rates T'_x, T'_y and T'_z has already been introduced in section 7.1.2: The datum transformation (7.2) is applied to the orbital state vectors of the slave acquisition by subtracting the relative motion $\vec{m} = (T_\mathrm{S} - T_\mathrm{M})\,\vec{v}$ of a representative ITRF station P_0 as depicted in figure 7.2b. The master orbit remains unchanged, because the master acquisition time T_M is assumed to be the reference epoch in which the target TRF coincides with the ITRF. This assumption is admissible, since the interferogram is insensitive to any constant translation that applies to both master and slave orbit.

The translation approach assumes that the motion of a tectonic plate is a pure translation in 3D space and can be approximated by the 3D velocity vector of a representative ITRF station. However, tectonic plates rather move along the curved earth's surface performing a spherical motion. Thus, a 3D velocity vector is only a local approximation but not appropriate to characterise the plate motion in a global sense. Due to the curvature of both earth and satellite orbit, its LOS component varies during the acquisition even though the vector itself remains constant in 3D space (see figure 7.5).

The approximation error of the correction approach thus causes a phase artefact with a dominant component in azimuth. It can be characterised by exploiting the equivalence of the reference frame effect with baseline errors:

$$\delta \dot{B}_\| = (T_\mathrm{S} - T_\mathrm{M})\,\dot{v}_\| \,\Delta t = B_T\,\Delta v_\| \,, \tag{7.6}$$

where $\Delta v_\|$ is the change of the LOS component of the assumed velocity vector $\vec{v}$ between early and late azimuth. With eq. (3.15) follows:

$$\Delta_\mathrm{az}\delta\phi = -\Delta_\mathrm{az}\delta\phi_\mathrm{ref} = \frac{4\pi}{\lambda}\,B_T\,\Delta v_\| \,, \tag{7.7}$$

and by analogy to eq. (7.5):

$$\Delta_\mathrm{az}\delta D' = -\dot{v}_\|\Delta t = -\Delta v_\| \,. \tag{7.8}$$

For Envisat it can be concluded that the bias in azimuth is generally much smaller than the reference frame effect in range as long as only a single SAR frame of 100 km length is processed (see figure 7.6). Hence, the benefit of the correction outweighs its model error.

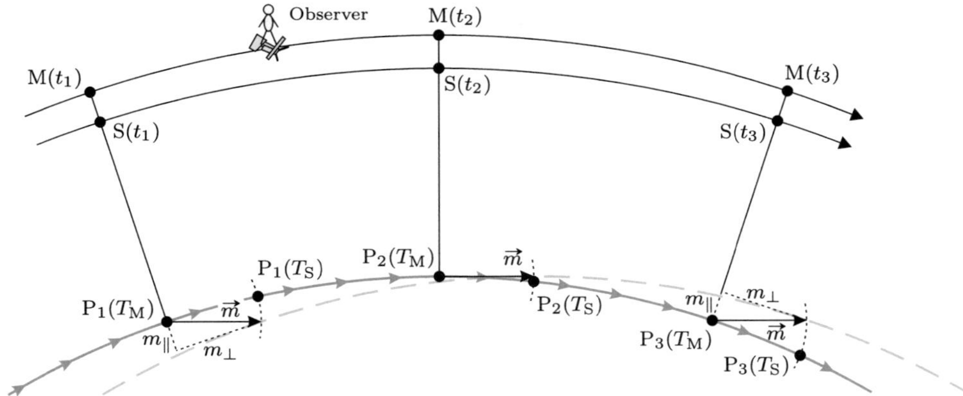

Figure 7.5.: Approximation of a purely horizontal plate motion by the 3D displacement vector $\vec{m} = (T_\mathrm{S} - T_\mathrm{M})\,\vec{v}$ in P_2 for different azimuth times t_i. The suggested rotation-free plate motion implies subsidence in P_1 and uplift in P_3, which does not comply with the actual movement following the curved surface.

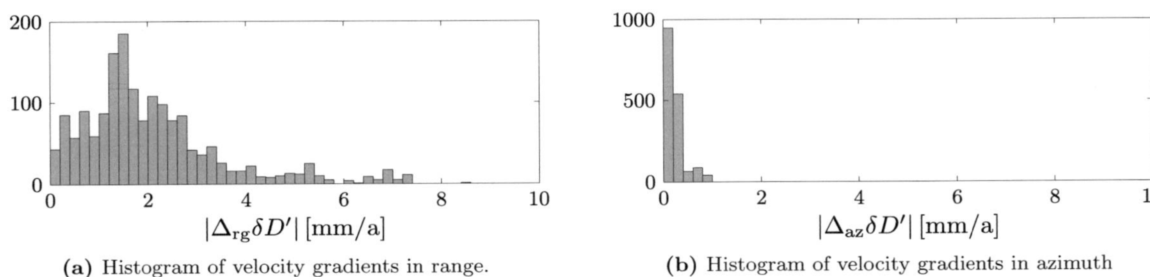

(a) Histogram of velocity gradients in range.　　(b) Histogram of velocity gradients in azimuth

Figure 7.6.: Global evaluation of trends that would result from a correction of the reference frame effect with the translation approach, predicted for full Envisat IS2 frames from one ascending and one descending track at 840 ITRF2008 stations. **(a)** Relative LOS velocity biases in range that are induced by the reference frame effect and can be removed with the translation approach. **(b)** Relative LOS velocity biases in azimuth that are induced due to approximation errors of the translation approach.

Representativity

If the area of interest is subject to significant tectonic deformation, it is impossible to find an ITRF station, the velocity of which is representative for the tectonic motion of the whole scene. But this is neither necessary nor expedient, since the correction does not intend to comprehensively anticipate the measured displacement field. It merely defines three translation rates of the geodetic datum, representing the assumed velocity of the reference point. Hence, the only requirement to be fulfilled is the representativity of the ITRF velocity for the reference point. Besides, it is self-evident that an ITRF velocity can only be considered representative if the orbital state vectors are expressed in the ITRF or an equivalent reference frame.

Due to successively evolving ITRF releases, another matter of representativity are the underlying ITRF releases of orbit data and correction approach. Although subsequent ITRF releases basically represent the same reference system, station positions and velocities are subject to small changes due to improved estimation strategies. Every time a new release is published, it has to be decided how to proceed with operationally processed orbit solutions. A switch in the processing strategy to the new frame may be considered as well as a complete reprocessing of older mission phases (M. Otten, ESOC, pers. comm., 2010). Keeping the old frame as reference would avoid discontinuities in the data, but on the other hand no benefit could be drawn from the enhanced frame consistency.

Table 7.3.: Estimated transformation parameters (translations T_x, T_y, T_z, differential scale D, rotation angles R_x, R_y, R_z and their rates of change) between recent ITRF realisations (ALTAMIMI ET AL., 2002, 2007, 2011). Scale and rotation parameters have been multiplied by a mean earth radius $R_E = 6371$ km to illustrate the impact on station coordinates. Note that there is no strict analytical relation between two ITRF releases; the estimated parameters rather provide a rough idea of the actual datum shift.

from	to	T_x	T_y	T_z	$R_E D$	$R_E R_x$	$R_E R_y$	$R_E R_z$	T'_x	T'_y	T'_z	$R_E D'$	$R_E R'_x$	$R_E R'_y$	$R_E R'_z$
					[mm]							[mm/a]			
ITRF2008	ITRF2005	-0.5	-0.9	-4.7	6.0	0.0	0.0	0.0	0.3	0.0	0.0	0.0	0.0	0.0	0.0
ITRF2005	ITRF2000	0.1	-0.8	-5.8	2.5	0.0	0.0	0.0	-0.2	0.1	-1.8	0.5	0.0	0.0	0.0
ITRF2000	ITRF97	6.7	6.1	-18.5	9.9	0.0	0.0	0.0	0.0	-0.6	-1.4	0.3	0.4	0.4	0.4

Even though it is advisable to always use homogeneous data sets for processing, the choice of the ITRF release affects orbit solutions only on the millimetre level. This can be seen from table 7.3, where estimated parameters of similarity transformations between recent ITRF solutions provide a rough idea of the actual datum shift. Only a translation in direction of the perpendicular baseline has a significant effect on InSAR processing. Even for the most pessimistic circumstances, where the perpendicular baseline is collinear with the z-axis of the global frame and the temporal baseline is very long, the effect on $\delta B_\perp$ cannot exceed a few centimetres according eq. (7.3). Hence, the choice of the correct ITRF release is not of primary importance for the correction of the reference frame effect.

7.2.2. Euler Rotation

Compared with three-dimensional velocity vectors, a better approximation of plate motion is provided by Euler vectors (see figure 7.7). An Euler vector points from the earth's centre to a specified Euler pole (Φ, Λ), and its modulus equals the angular velocity ω with which the tectonic plate performs a rotation about this pole. Thus, any spherical motion can be described globally for rigid plates by three parameters. Only atypical movements like plate-wide subsidence, uplift or tilting cannot be accommodated.

Euler vectors for the individual plates are available from dedicated plate kinematic models. If orbital state vectors are expressed in the ITRF and the Euler vector parameters (Φ, Λ, ω) are available in the

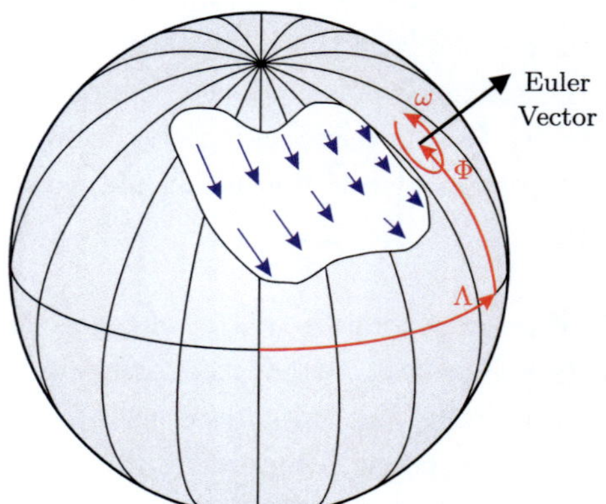

Figure 7.7.: Euler rotation of a tectonic plate about its Euler pole. The plate motion is quantified by the geographic coordinates (Φ, Λ) of the Euler pole and the angular velocity ω. The characterising Euler vector is pointing from the earth's centre to the Euler pole and has the modulus ω.

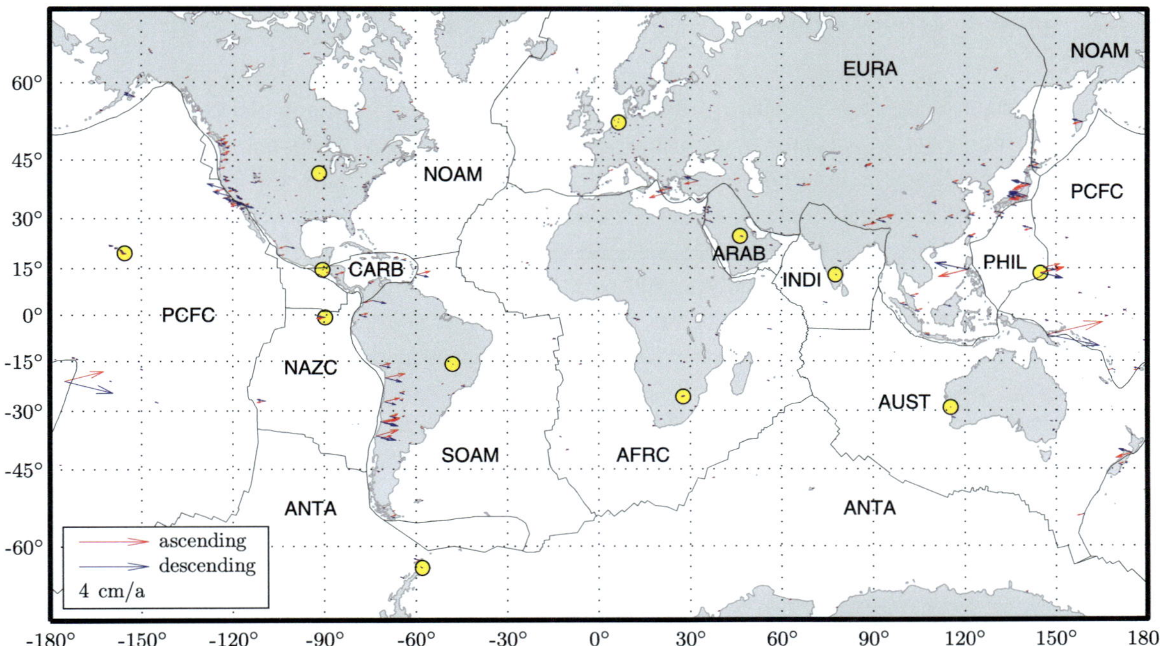

Figure 7.8.: Bias of the baseline error rates $\delta B'_{\perp} = v_{\perp}$ due to approximation of the ITRF2008 velocities by point velocities computed from NNR-NUVEL-1A, evaluated for Envisat IS2 interferograms at 840 ITRF2008 stations. The arrows are aligned to the horizontal projection of the respective radar line of sight. Mind the enlarged scale with respect to figure 7.3. Numerical values for the twelve encircled stations can be deduced from table 7.1.

same datum, an applicable datum transformation to correct for the reference frame effect can be formulated by analogy to eq. (7.2):

$$\vec{x}_{\text{plate}}(T) = \vec{x}_{\text{ITRF}} - (T - T_0) \left[\omega \begin{pmatrix} \cos\Phi\cos\Lambda \\ \cos\Phi\sin\Lambda \\ \sin\Phi \end{pmatrix} \times \vec{x}_{\text{ITRF}} \right]. \tag{7.9}$$

This approach has the advantage that the approximation quality does not depend on the availability of an ITRF station close to the reference point for InSAR velocities. Furthermore, the correction is less inclined to produce phase artefacts than the translation approach, since it focusses on the dominant horizontal components of plate motion under consideration of the earth's curvature. However, plate kinematic models are sometimes partly or fully based on geological data reflecting the average motion from a few million years ago to date, which is not necessarily representative for the recent era of remote sensing. Due to plate-internal deformation, it may also occur that an Euler vector adjusted to the whole plate does not adequately describe the motion of a particular reference point.

In any case, it is important to assure that the Euler vectors are given in (or transformed to) the same datum as the orbit reference frame, i. e., the ITRF. This requirement is definitely satisfied for the purely geological model NNR-NUVEL-1A (DeMets et al., 1994), because the ITRF2000 and later releases have adopted its datum (see section 7.1.1). Figure 7.8 shows the biases of the baseline error rates at the ITRF2008 stations that occur if the reference frame effect is corrected using NNR-NUVEL-1A Euler vectors. Except for sites located in the vicinity of plate boundaries, the deviation is in the order of a few mm/a and thus insignificant. An only marginally better approximation was obtained with models based on geodetic observations only (Sella et al., 2002; Drewes, 2009).

7.2.3. General Transformation

The two approaches discussed so far consider only three degrees of freedom, whereas a most general correction would involve a datum transformation with six degrees of freedom. Such a procedure is reasonable for applications in which a local GNSS velocity field is available and InSAR measurements are used to densify this field. In this case, there is not one reference point with an assumed velocity of zero but rather a whole set of reference stations with individual velocities. These reference stations are required to represent the assumedly stable part of the area of interest, and their coordinates and velocities must be given in a unique reference frame. Evidently, neither the velocity of a single station nor the three parameters of an Euler vector can adequately represent the datum of a GNSS network.

A most general correction approach, guaranteeing that the measured InSAR velocities are obtained in the same TRF as the (GNSS) reference stations, may involve the following steps:

1. If the velocities $\dot{\vec{y}}_i$ of the reference stations $\vec{y}_i = (x_i, y_i, z_i)^T$ $(i = 1 \ldots n)$ are not already given in the ITRF, they have to be transformed from their inherent "TRF0" to the ITRF. If these two frames are close to each other, a simplified transformation with differential rotations may be used:

$$\dot{\vec{y}}_{i,\mathrm{ITRF}} = \dot{\vec{y}}_{i,\mathrm{TRF0}} + \begin{pmatrix} T'_{x,0} \\ T'_{y,0} \\ T'_{z,0} \end{pmatrix} + \begin{pmatrix} 0 & R'_{z,0} & -R'_{y,0} \\ -R'_{z,0} & 0 & R'_{x,0} \\ R'_{y,0} & -R'_{x,0} & 0 \end{pmatrix} \vec{y}_i \,. \tag{7.10}$$

$T'_{x,0}$, $T'_{y,0}$, $T'_{z,0}$, $R'_{x,0}$, $R'_{y,0}$ and $R'_{z,0}$ are six transformation parameters that have to be estimated from identical points. Considering that the equation system (7.10) has only a rank of five for $n = 2$, at least three stations with given ITRF velocities are required in the reference network.

2. Three translation rates (T'_x, T'_y, T'_z) and three rotation rates (R'_x, R'_y, R'_z) of the reference network are estimated from the given ITRF velocities $\dot{\vec{y}}_{i,ITRF}$ for all n reference stations. The associated overdetermined system of $3n$ observation equations reads:

$$\mathrm{E}\left\{\dot{\vec{y}}_{i,\mathrm{ITRF}}\right\} = \begin{pmatrix} T'_x \\ T'_y \\ T'_z \end{pmatrix} + \begin{pmatrix} 0 & R'_z & -R'_y \\ -R'_z & 0 & R'_x \\ R'_y & -R'_x & 0 \end{pmatrix} \vec{y}_i \,, \quad i = 1 \ldots n \,. \tag{7.11}$$

3. The estimated mean translational and rotational motion of the reference network is inversely applied to the orbit positions $\vec{x}$ by analogy to eq. (7.2):

$$\vec{x}_{\mathrm{plate}}(T) = \vec{x}_{\mathrm{ITRF}} - (T - T_0) \left[\begin{pmatrix} T'_x \\ T'_y \\ T'_z \end{pmatrix} + \begin{pmatrix} 0 & R'_z & -R'_y \\ -R'_z & 0 & R'_x \\ R'_y & -R'_x & 0 \end{pmatrix} \vec{x}_{\mathrm{ITRF}} \right] \,. \tag{7.12}$$

Conveniently choosing $T_0 := T_\mathrm{M}$ implies that the master orbit remains unchanged and that the update to the slave orbit increases linearly with the temporal baseline.

Applied to the orbital state vectors, the datum transformation (7.12) ensures a reference phase that is free of artefacts due to reference frame inconsistencies. The reduced interferometric phases φ are still relative quantities, from which absolute LOS velocities can only be inferred if at least one reference point with a given velocity is available. In case of a GNSS network, there are multiple reference points, and the most self-evident strategy to obtain InSAR velocities is to minimise their deviation from the GNSS velocity components in the InSAR line of sight at all these points.

7.3. Conclusions

Neglecting relative motion between the earth's surface and the coordinate frame in which satellite orbits are expressed induces a trend into the interferograms that is almost linear in range and can bias Envisat displacement rates by up to 7 mm/a. The bias of an estimated relative displacement depends on the geographical location as well as on the separation of two measurements in both time and range. In contrast to orbital ramps, this *reference frame effect* is correlated in time and thus not separable from an actual deformation signal.

Being negligible for local phenomena, correcting for the reference frame effect should be considered whenever large-scale deformation signals are the subject of InSAR analysis. For instance, the measurement of tectonic strain can be significantly biased if no dedicated correction is applied. The bias may become more than five times the desirable accuracy, which was amounted by WRIGHT ET AL. (2011) to a LOS velocity gradient of 1.2 mm/a over a ground distance of 100 km.

Whereas the predictions for a correction with the translation approach have already been validated for a PS-InSAR time series (BÄHR ET AL., 2012), the reference frame effect itself has not been explicitly observed yet. Nevertheless, its existence is evident without explicit verification, and a correction is advisable whenever large-scale deformation phenomena are measured with high accuracy requirements. Based on the comparison in figure 7.8, a dedicated Euler rotation of orbital state vectors can be considered sufficiently accurate in most cases. Also a homogeneous translation derived from one representative ITRF velocity performs well but may involve minor phase artefacts. It should be preferred nevertheless in regions where the approximation quality of plate kinematic models is bad. The general transformation approach may be worth considering whenever InSAR measurements are used to densify an existing (GNSS) velocity field. If the benefit of the complex transformation outweighs the involved effort in this case still needs to be investigated.

InSAR deformation estimates are not necessarily biased if the reference frame effect is neglected. The adequacy of a dedicated correction depends on the implied notion of reference. If a deformation map is supposed to display the relative movement with respect to the centre of mass of the earth and the conventional global coordinate reference system, no correction needs to be applied, because the satellite orbits are already given in the appropriate datum. However, usually the measured deformation is conventionally interpreted relative to an assumedly stable part of the earth's surface. Then, a correction is indispensable unless it can be shown that the induced bias, which depends on the respective location on earth and the spatial extension of the area of interest, is insignificant.

8. Conclusions and Outlook

Concerned with the impact of inaccurate satellite orbits on SAR interferometry, this thesis provides a detailed description of error mechanisms, comprising an evaluation of their significance and proposing applicable correction approaches. This final chapter summarises the detailed conclusions and recommendations from previous chapters in order to compose a general picture of the relevance of orbital effects for present and future SAR missions.

Recent advancements in precise orbit determination have confined the impact of baseline errors due to orbit inaccuracies below the significance threshold for most acquisitions of modern sensors. Although it can be expected that orbit inaccuracies will be negligible for most InSAR applications of future missions, their insignificance can never be guaranteed. Their relevance will persist for acquisitions of older sensors, processing historical data, occasional shortcomings in orbit accuracy, large-scale deformation analyses with high performance requirements, single interferograms, short time series, near real-time applications and quality assessment.

In order to estimate and subsequently correct for the effect of baseline errors, a linear phase ramp is an acceptable approximation for scenes with small or moderate elevation differences. A slightly more accurate and elevation-insensitive parameterisation comprises the error $\delta B_\perp$ in the perpendicular baseline and the error $\delta \dot{B}_\parallel$ in the rate of change of the parallel baseline. Both representations enable a stable solution to the inverse problem of inferring baseline errors from the residual interferometric phase. Adjusting the relative baseline errors in a redundant network of linearly dependent interferometric combinations does not only yield quasi-absolute, acquisition-wise orbit errors but also provides a powerful instrument of quality control. Thus, inconsistencies of the least squares baseline error estimates due to unwrapping errors can be reliably identified. This is different for the alternatively proposed gridsearch approach, which yields partly inconsistent estimates and fails in inferring blunders from network misclosures.

The proposed methodology still leaves some room for improvement. Especially the failure of outlier detection among gridsearch estimates may be remedied by an optimised weighting scheme. Furthermore, the stochastic model of the least squares estimator is still deficient in several respects. However, the success of developing a more adequate model with a significantly better performance is questionable, as the achievable benefit might not outweigh the invested effort. Concerning the functional model of both estimators, further development under consideration of robust estimation techniques may mitigate the sensitivity of estimates to the spatial distribution of observations. Nonetheless, the most efficient strategy to further enhance the presented estimation techniques is to overcome model deficiencies by engaging an integration into comprehensive approaches that involve a likewise rigorous modelling for other signal contributions. Particular benefit may be drawn from the integration of numerical weather models.

There are two basic scenarios in which orbit error estimation can support InSAR time series analyses. On the one hand, a priori estimation and removal of the orbital contribution can support phase unwrapping if orbit errors are relatively large as it is often the case for the Radarsat-1/2 missions. This pre-processing application is limited to the gridsearch estimator that can process the wrapped phase. On the other hand, orbit error estimation with any estimator can be integrated into the processing chain at a later stage, either jointly with other contributions or iteratively alternating. Thus, explicitly estimating orbital and atmospheric signal components helps isolating the deformation signal, whereas an explicit distinction between orbital and atmospheric effects is not necessary. The algorithms proposed in this thesis can be considered a contribution to a comprehensive methodology.

Besides baseline errors, a number of less relevant but still considerable error mechanisms have been investigated. One of them is related to absolute errors in the annotated timing of SAR images. These are occasionally large enough to induce significant phase ramps but can also be estimated with sufficient accuracy by image correlation. Another potential cause of ramp-like error signals is a continuous drift of the radar frequency, which is difficult to validate. Nevertheless, there is some evidence that all time series of Envisat data are affected by a frequency drift that is large enough to become significant for interferometric applications. Small but significant phase ramps may also result from neglecting relative motion of tectonic plates and the orbit reference frame. This *reference frame effect* can easily be accounted for by one of three proposed correction approaches.

A. Model Conventions and Simulations

Many simulations in chapters 3 and 4 are based on a consistent geometric model of a SAR acquisition. The intention of this appendix is to make all model conventions transparent in order to enable the reader to reproduce the presented numbers, fringe patterns and coregistration offset patterns. Table A.1 collects some basic physical and model constants.

Table A.1.: Physical and model constants used for simulations (c and GM from Montenbruck and Gill, 2000).

Speed of light	c	$299792458\,\mathrm{ms^{-1}}$
Geocentric gravitational constant	GM	$398600.4415\,\mathrm{km^3s^{-2}}$
Mean earth radius	R_E	$6371\,\mathrm{km}$

Almost all simulations assume a spherical satellite orbit about a concentric spherical earth with a constant radius R_E (see figure A.1a). Only the fringe pattern in figure 3.8a is exceptionally based on an ellipsoidal reference surface. For the Taylor series in section 3.4, the partial derivatives of the master look angle θ with respect to the range $R := R_\mathrm{M}$ in the spherical model are of special interest:

$$
\begin{aligned}
\frac{\partial \theta}{\partial R}(R) &= \frac{H(2R_E + H) - R^2}{2R^2(R_E + H)\sin\theta(R)} = -\frac{R_\mathrm{E}\cos\theta_\mathrm{inc}}{R(R_E + H)\sin\theta(R)} \\
\frac{\partial^2 \theta}{\partial R^2}(R) &= -\cot\theta(R)\left(\frac{\partial\theta}{\partial R}(R)\right)^2 - \frac{H(2R_E + H)}{R^3(R_E + H)\sin\theta(R)} \, .
\end{aligned}
\tag{A.1}
$$

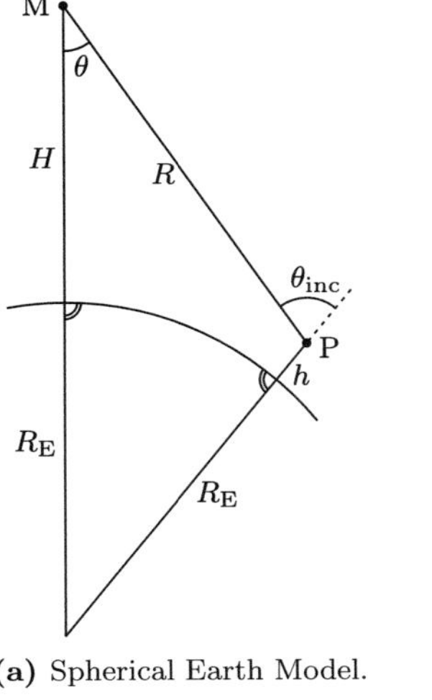

(a) Spherical Earth Model.

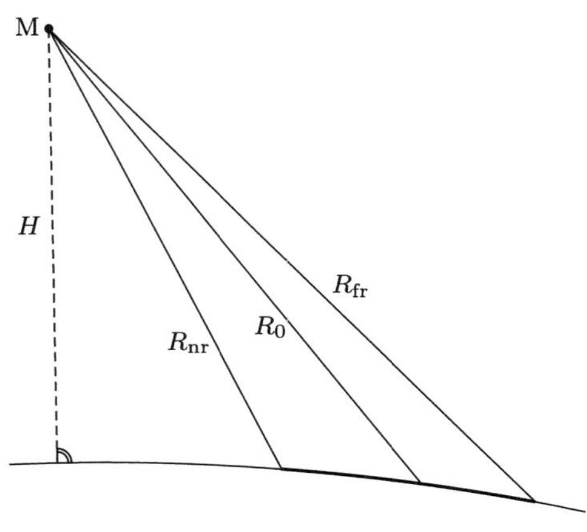

(b) Near range (nr) and far range (fr).

Figure A.1.: Acquisition geometry that is assumed for most numerical simulations. **(a)** The earth body is approximated by a sphere with a radius of $R_\mathrm{E} = 6371\,\mathrm{km}$. **(b)** The acquired swath extends from near range $R_\mathrm{nr} = \frac{c}{2}\tau_0$ to far range $R_\mathrm{fr} = \frac{c}{2}(\tau_0 + \frac{n_\eta}{f_\mathrm{RSR}})$, where n_η is the number of samples in range. Note that the average range R_0 as defined in eq. (A.5) is *not* bisecting the sensor's field of view.

For the master incidence angle θ_{inc} follows:

$$\frac{\partial \theta_{\mathrm{inc}}}{\partial R} = \frac{R^2 + H(2R_{\mathrm{E}} + H)}{2R^2 R_{\mathrm{E}} \sin \theta_{\mathrm{inc}}} = -\frac{(R_{\mathrm{E}} + H) \cos \theta}{R R_{\mathrm{E}} \sin \theta_{\mathrm{inc}}} \; . \tag{A.2}$$

As stated in the introduction to chapter 3, all basic simulations apply to Envisat's Image Swath 2 (IS2) and are thus approximately conferrable to the ERS geometry. Beam modes of some other sensors are considered within the scope of this appendix (see table 3.1 for an overview of selected spaceborne sensors), limited either to the unique conventional stripmap beam (if available) or to all beams of the respective standard stripmap mode. For the Radarsat satellites, the selection has been additionally extended to fine beam modes, which make an important contribution to the interferometric capability of Radarsat-1 (VACHON ET AL., 1995). Tables A.2 through A.4 provide some characteristic numbers supplementing considerations in chapters 3 and 4.

For some variable parameters, appropriate conventions have been made. As the satellite altitude H varies during an elliptic orbit above an approximately ellipsoidal earth, either a mean altitude or the altitude at the equator has been adopted (see table 3.1), depending on the availability of official specifications. The velocity of the platform is computed for a circularly orbiting body about a radially layered spherical earth:

$$v = \sqrt{\frac{GM}{R_{\mathrm{E}} + H}} \; . \tag{A.3}$$

The total acquisition time is computed from the scene length L by rule of proportion:

$$\Delta t = \frac{L}{v} \cdot \frac{R_{\mathrm{E}} + H}{R_{\mathrm{E}}} \; . \tag{A.4}$$

Some computations require a mean value for range-dependent parameters like θ, θ_{inc} or the range R itself. They refer to the pixel at mid-range R_0, which is defined as the arithmetic mean of near range and far range (see figure A.1b):

$$R_0 = \frac{R_{\mathrm{nr}} + R_{\mathrm{fr}}}{2} \; . \tag{A.5}$$

Complementarily, the definition of the overall range span reads:

$$\Delta R = R_{\mathrm{fr}} - R_{\mathrm{nr}} \; . \tag{A.6}$$

Table A.2.: Parameters of selected acquisition modes of the selection of sensors made in table 3.1: look angle $\theta(R_\mathrm{nr})$, $\theta(R_0)$, $\theta(R_\mathrm{fr})$, incidence angle $\theta(R_\mathrm{nr})$, $\theta(R_0)$, $\theta(R_\mathrm{fr})$, mid-range $R_0 = (R_\mathrm{nr}+R_\mathrm{fr})/2$, scene length L, swath width W, acquisition time Δt and range span $\Delta R = R_\mathrm{fr} - R_\mathrm{nr}$. The geometry of a mode is defined by the three bold parameters that have been taken from nominal specifications where available (DLR, 2009; ESA, 2007a,b; JAXA, 2008; GEUDTNER ET AL., 2011; MDA, 2009; MIRANDA AND ROSICH, 2011; RSI, 2000). All remaining parameters are computed as complements to a consistent geometry. Note that the numbers do not apply precisely to any individual data product and may vary slightly as the case arises. Small deviations from table 1 in (BÄHR AND HANSSEN, 2012) result from different assumptions of system parameters.

Sensor	Mode	θ [°]	θ_inc [°]	R_0 [km]	L [km]×[km] W	Δt [s]	ΔR [km]
ALOS	FBS7	**31.6**...**34.4**...**36.8**	35.6...38.8... 41.7	871	**70**× 101	10	63
ERS-1/2		**16.9**...20.3... 23.1	19.1...23.0... 26.2	850	**100**×**100**	15	39
Envisat	IS2	**16.9**...20.3... 23.1	19.1...23.0... 26.2	850	**100**×**100**	15	39
Radarsat-1	F1	32.2...33.5... 34.7	**36.9**...38.4...**39.9**	988	**50**× 55	8	34
	F2	34.2...35.4... 36.6	**39.3**...40.7...**42.1**	1015	**50**× 54	8	35
	F3	36.0...37.1... 38.2	**41.4**...42.8...**44.1**	1042	**50**× 55	8	37
	F4	37.5...38.6... 39.6	**43.3**...44.6...**45.9**	1069	**50**× 55	8	39
	F5	39.1...40.1... 41.1	**45.2**...46.5...**47.7**	1098	**50**× 56	8	40
	S1	17.7...21.1... 24.0	**20.0**...23.9...**27.2**	866	**100**× 103	15	41
	S2	21.2...24.6... 27.5	**24.0**...28.0...**31.3**	892	**100**× 111	15	51
	S3	26.7...29.7... 32.3	**30.4**...33.9...**37.0**	941	**100**× 111	15	61
	S4	29.4...32.2... 34.6	**33.5**...36.8...**39.7**	970	**100**× 110	15	66
	S5	31.7...34.4... 36.6	**36.3**...39.4...**42.2**	999	**100**× 110	15	70
	S6	35.8...38.1... 40.1	**41.2**...44.0...**46.5**	1059	**100**× 110	15	77
	S7	38.4...40.5... 42.3	**44.4**...47.0...**49.3**	1106	**100**× 110	15	80
Radarsat-2	F1	32.2...33.5... 34.7	**36.9**...38.4...**39.9**	988	**50**× 55	8	34
	F2	34.2...35.4... 36.6	**39.3**...40.7...**42.1**	1015	**50**× 54	8	35
	F3	36.0...37.1... 38.2	**41.4**...42.8...**44.1**	1042	**50**× 55	8	37
	F4	37.5...38.6... 39.6	**43.3**...44.6...**45.9**	1069	**50**× 55	8	39
	F5	39.1...40.1... 41.1	**45.2**...46.5...**47.7**	1098	**50**× 56	8	40
	S1	17.7...21.1... 24.0	**20.0**...23.9...**27.2**	866	**100**× 103	15	41
	S2	21.2...24.6... 27.5	**24.0**...28.0...**31.3**	892	**100**× 111	15	51
	S3	26.7...29.7... 32.3	**30.4**...33.9...**37.0**	941	**100**× 111	15	61
	S4	29.4...32.2... 34.6	**33.5**...36.8...**39.7**	970	**100**× 110	15	66
	S5	31.7...34.4... 36.6	**36.3**...39.4...**42.2**	999	**100**× 110	15	70
	S6	35.8...38.1... 40.1	**41.2**...44.0...**46.5**	1059	**100**× 110	15	77
	S7	38.4...40.5... 42.3	**44.4**...47.0...**49.3**	1106	**100**× 110	15	80
Sentinel-1	IWS	22.4...31.0... 36.8	**25.0**...34.9... 41.6	833	**170**×**250**	25	138
TerraSAR-X	strip_003	**18.2**...**19.8**...**21.3**	19.7...21.5... 23.1	549	**50**× 32	7	12
	strip_004	**20.6**...22.1...**23.5**	22.3...24.0... 25.6	559	**50**× 32	7	13
	strip_005	**22.8**...24.3...**25.7**	24.8...26.4... 28.0	569	**50**× 32	7	14
	strip_006	**25.0**...26.5...**27.8**	27.2...28.8... 30.3	580	**50**× 32	7	15
	strip_007	**27.1**...28.5...**29.8**	29.5...31.1... 32.5	592	**50**× 32	7	16
	strip_008	**29.2**...30.5...**31.7**	31.8...33.2... 34.6	605	**50**× 32	7	17
	strip_009	**31.1**...32.3...**33.5**	33.9...35.3... 36.6	619	**50**× 32	7	18
	strip_010	**32.9**...34.1...**35.2**	36.0...37.3... 38.6	633	**50**× 32	7	19
	strip_011	**34.7**...35.8...**36.9**	38.0...39.2... 40.4	648	**50**× 32	7	20
	strip_012	**36.4**...37.4...**38.4**	39.8...41.1... 42.2	663	**50**× 32	7	21
	strip_013	**38.0**...39.0...**39.9**	41.7...42.8... 43.9	680	**50**× 32	7	22
	strip_014	**39.5**...40.4...**41.3**	43.4...44.5... 45.5	696	**50**× 32	7	22

Table A.3.: Sensitivity of some parameters with respect to changes in the interferometric phase. The numbers indicate how much a parameter has to be changed to induce an error signal spanning 2π between extreme values. Evaluations for different sensors follow the specifications in table A.2. The baseline errors $\delta B_{\parallel,2\pi}$, $\delta B_{\perp,2\pi}$, $\delta\dot{B}_{\parallel,2\pi}$ and $\delta\dot{B}_{\perp,2\pi}$ are defined in eq. (3.17). The timing error $\delta\tau_{\mathrm{M},2\pi}$ is defined in eq. (3.22), computed for $B_{\parallel} = 0\,\mathrm{m}$ and inversly proportional to $B_{\perp}$. The clock errors $\Delta\nu_{2\pi}$ and $\Delta\dot{\nu}_{2\pi}$ are defined in eq. (3.27).

Sensor	Mode	$\delta B_{\parallel,2\pi}$ [m]	$\delta B_{\perp,2\pi}$ [m]	$\delta\dot{B}_{\parallel,2\pi}$ [mm/s]	$\delta\dot{B}_{\perp,2\pi}$ [mm/s]	$\delta\tau_{\mathrm{M},2\pi}$ [μs] $B_{\perp}=$ 100 m	$\delta\tau_{\mathrm{M},2\pi}$ [μs] $B_{\perp}=$ 500 m	$\Delta\nu_{2\pi}$ [ppm]	$\Delta\dot{\nu}_{2\pi}$ [ppb/s]
ALOS	FBS7	-114.6	1.30	11.4	251	19.1	3.81	1.87	13.1
ERS-1/2		-19.2	0.26	1.9	35	1.8	0.35	0.73	2.2
Envisat	IS2	-19.0	0.26	1.9	34	1.7	0.35	0.73	2.2
Radarsat-1	F1	-118.4	0.65	3.7	171	10.4	2.08	0.83	3.8
	F2	-138.8	0.70	3.7	186	12.4	2.47	0.80	3.7
	F3	-152.3	0.73	3.7	194	14.0	2.80	0.76	3.6
	F4	-167.8	0.77	3.7	204	15.7	3.14	0.73	3.5
	F5	-185.7	0.81	3.7	215	17.6	3.52	0.70	3.4
	S1	-18.9	0.26	1.9	34	1.9	0.38	0.68	2.2
	S2	-18.7	0.26	1.9	34	2.5	0.49	0.55	2.1
	S3	-23.6	0.29	1.9	38	3.8	0.77	0.46	2.0
	S4	-27.3	0.31	1.9	41	4.7	0.94	0.43	1.9
	S5	-30.9	0.33	1.9	44	5.6	1.11	0.40	1.9
	S6	-40.0	0.38	1.9	50	7.5	1.50	0.37	1.8
	S7	-48.6	0.41	1.9	55	9.2	1.84	0.35	1.7
Radarsat-2	F1	-116.1	0.63	3.7	168	10.2	2.04	0.81	3.7
	F2	-136.1	0.69	3.7	182	12.1	2.43	0.79	3.6
	F3	-149.4	0.72	3.7	191	13.7	2.75	0.75	3.5
	F4	-164.5	0.76	3.7	200	15.4	3.08	0.72	3.4
	F5	-182.1	0.79	3.7	211	17.2	3.45	0.69	3.3
	S1	-18.5	0.25	1.8	34	1.9	0.37	0.67	2.1
	S2	-18.3	0.25	1.8	33	2.4	0.48	0.54	2.1
	S3	-23.2	0.28	1.8	38	3.8	0.75	0.45	2.0
	S4	-26.8	0.30	1.8	40	4.6	0.92	0.42	1.9
	S5	-30.3	0.32	1.8	43	5.5	1.09	0.40	1.8
	S6	-39.3	0.37	1.8	49	7.4	1.48	0.36	1.7
	S7	-47.7	0.41	1.8	54	9.0	1.80	0.34	1.7
Sentinel-1	IWS	-3.5	0.11	1.1	9	1.4	0.28	0.20	1.3
TerraSAR-X	strip_003	-42.7	0.29	2.2	81	1.2	0.23	1.34	4.0
	strip_004	-45.8	0.30	2.2	84	1.5	0.29	1.20	3.9
	strip_005	-49.4	0.31	2.2	87	1.8	0.36	1.10	3.8
	strip_006	-53.6	0.32	2.2	91	2.2	0.43	1.01	3.8
	strip_007	-58.4	0.34	2.2	95	2.6	0.52	0.95	3.7
	strip_008	-64.0	0.35	2.2	99	3.0	0.61	0.89	3.6
	strip_009	-70.3	0.37	2.2	104	3.5	0.70	0.84	3.5
	strip_010	-77.5	0.39	2.2	109	4.0	0.81	0.80	3.5
	strip_011	-85.6	0.41	2.2	115	4.6	0.92	0.77	3.4
	strip_012	-94.7	0.43	2.2	121	5.2	1.04	0.74	3.3
	strip_013	-105.0	0.45	2.2	127	5.9	1.18	0.72	3.2
	strip_014	-116.4	0.48	2.2	134	6.6	1.32	0.70	3.1

Table A.4.: Maximum deformation bias $\Delta\delta D$ (see eq. (4.23)) due to approximation of the orbital error signal by a dedicated parametric model, evaluated for different sensors following the specifications in table A.2. Simulations are based on an error baseline of $\delta B = 1\,\mathrm{dm}$ length with varying orientation β that is constant over azimuth, where the indicated $\Delta\delta D$ are maximised with respect to β. p_0 stands for the uncorrected orbital error signal, which can also be interpreted as error of an approximation with a zero-degree polynomial (see figure 4.6a). Other considered models are a parameterisation by the baseline error components $(\delta\dot{B}_\parallel, \delta B_\perp)$ and approximations by a linear ramp $p_1(\xi,\eta)$ or a paraboloid $p_2(\xi,\eta)$, respectively (see figure 4.6b). Biases have been computed for a flat earth and a pyramidal evaluation model (see figure 4.7a). Deviations from table 1 in (Bähr and Hanssen, 2012) can be explained by slightly different assumptions of system parameters.

Sensor	Mode	$\max_\beta(\Delta\delta D)$ [mm]						
			Flat Earth			Pyramid, $h_{\max} = 1\,\mathrm{km}$		
		p_0	$(\delta\dot{B}_\parallel, \delta B_\perp)$	$p_1(\xi,\eta)$	$p_2(\xi,\eta)$	$(\delta\dot{B}_\parallel, \delta B_\perp)$	$p_1(\xi,\eta)$	$p_2(\xi,\eta)$
ALOS	FBS7	9.1	0.11	0.36	0.03	0.11	0.54	0.06
Envisat	IS2	10.9	0.16	0.59	0.07	0.16	0.89	0.12
ERS-1/2		10.9	0.16	0.59	0.07	0.16	0.89	0.12
Radarsat-1/2	F1	4.4	0.02	0.08	0.00	0.02	0.24	0.04
	F2	4.0	0.02	0.07	0.00	0.02	0.22	0.04
	F3	3.9	0.02	0.07	0.00	0.02	0.21	0.04
	F4	3.7	0.02	0.06	0.00	0.02	0.19	0.04
	F5	3.5	0.02	0.06	0.00	0.02	0.18	0.03
	S1	10.9	0.16	0.59	0.07	0.16	0.87	0.12
	S2	11.0	0.16	0.55	0.06	0.16	0.79	0.10
	S3	9.8	0.12	0.42	0.03	0.13	0.61	0.07
	S4	9.1	0.11	0.36	0.03	0.11	0.53	0.06
	S5	8.6	0.10	0.32	0.02	0.10	0.48	0.05
	S6	7.5	0.07	0.26	0.02	0.07	0.39	0.04
	S7	6.8	0.06	0.22	0.01	0.06	0.34	0.04
Sentinel-1	IWS	25.1	0.88	2.87	0.57	0.88	3.08	0.60
TerraSAR-X	strip_003	5.4	0.04	0.14	0.01	0.04	0.64	0.14
	strip_004	5.2	0.03	0.12	0.01	0.04	0.57	0.12
	strip_005	5.0	0.03	0.11	0.01	0.03	0.51	0.11
	strip_006	4.8	0.03	0.10	0.00	0.03	0.46	0.10
	strip_007	4.6	0.03	0.09	0.00	0.03	0.42	0.09
	strip_008	4.4	0.02	0.08	0.00	0.03	0.38	0.08
	strip_009	4.2	0.02	0.07	0.00	0.02	0.35	0.08
	strip_010	4.0	0.02	0.07	0.00	0.02	0.33	0.07
	strip_011	3.8	0.02	0.06	0.00	0.02	0.30	0.07
	strip_012	3.6	0.02	0.06	0.00	0.02	0.28	0.06
	strip_013	3.4	0.02	0.05	0.00	0.02	0.27	0.06
	strip_014	3.3	0.01	0.05	0.00	0.01	0.25	0.06

Bibliography

ADAM N, BM KAMPES, M EINEDER, J WORAWATTANAMATEEKUL, M KIRCHER (2003): The Development of a Scientific Permanent Scatterer System. In: SCHROEDER M, K JACOBSEN, C HEIPKE (Eds.), *Proceedings of the Joint ISPRS/EARSeL Workshop "High Resolution Mapping from Space 2003", Hannover, Germany, 6–8 Oct 2003.*

ALTAMIMI Z, X COLLILIEUX, B GARAYT, J LEGRAND, C BOUCHER (2007): ITRF2005: A new release of the International Terrestrial Reference Frame based on time series of station positions and Earth Orientation Parameters. *Journal of Geophysical Research 112*:B09401. DOI: 10.1029/2007JB004949.

ALTAMIMI Z, X COLLILIEUX, L MÉTIVIER (2011): ITRF2008: an improved solution of the international terrestrial reference frame. *Journal of Geodesy 85*(8):457–473. DOI: 10.1007/s00190-011-0444-4.

ALTAMIMI Z, P SILLARD, C BOUCHER (2002): ITRF2000: A new release of the International Terrestrial Reference Frame for earth science applications. *Journal of Geophysical Research 107*(B10):2214. DOI: 10.1029/2001JB000561.

AMIRI-SIMKOOEI A (2007): *Least-Squares Variance Component Estimation.* Publications on Geodesy 64, NCG, Delft, The Netherlands.

ARIKAN M, FJ VAN LEIJEN, L GUANG, RF HANSSEN (2008): Improved Image Alignment under the Influence of Elevation. In: LACOSTE H, L OUWEHAND (Eds.), *Proceedings of 'FRINGE 2007', Frascati, Italy, 26–30 Nov 2007*, ESA SP-649.

ASI (2007): *COSMO-SkyMed System Description & User Guide.* Technical document no. ASI-CSM-ENG-RS-093-A, ASI.

BAARDA W (1968): *A testing procedure for use in geodetic networks.* Publications on Geodesy 9, NCG, Delft, The Netherlands.

BÄHR H, Z ALTAMIMI, B HECK (2007): *Variance Component Estimation for Combination of Terrestrial Reference Frames*, Schriftenreihe des Studiengangs Geodäsie und Geoinformatik, Vol. 2007,6. Universitätsverlag Karlsruhe, Karlsruhe, Germany.

BÄHR H, RF HANSSEN (2010): Network Adjustment of Orbit Errors in SAR Interferometry. In: LACOSTE-FRANCIS H (Ed.), *Proceedings of 'Fringe 2009', Frascati, Italy, 30 Nov–4 Dec 2009*, ESA SP-677.

BÄHR H, RF HANSSEN (2012): Reliable estimation of orbit errors in spaceborne SAR interferometry. The network approach. *Journal of Geodesy 86*(12):1147–1164. DOI: 10.1007/s00190-012-0571-6.

BÄHR H, S SAMIEI-ESFAHANY, RF HANSSEN (2012): On the Effect of Reference Frame Motion on InSAR Deformation Estimates. In: OUWEHAND L (Ed.), *Proceedings of 'Fringe 2011', Frascati, Italy, 19–23 Sep 2011*, ESA SP-697.

BAMLER R, P HARTL (1998): Synthetic aperture radar interferometry. *Inverse Problems 14*(4):R1–R54.

BAMLER R, B SCHÄTTLER (1993): SAR Data Acquisition and Image Formation. In: SCHREIER G (Ed.), *SAR geocoding: data and systems*, chapter 3, 53–102. Wichmann, Karlsruhe, Germany.

BEDRICH S (1998): Signal design and in-orbit performance of the new microwave satellite tracking system PRARE. In: *Proceedings of the 5th International Symposium on Spread Spectrum Techniques and Applications 1998, Sun City, South Africa, 2–4 Sep 1998.*

BERARDINO P, G FORNARO, R LANARI, E SANSOSTI (2002): A New Algorithm for Surface Deformation Monitoring Based on Small Baseline Differential SAR Interferograms. *IEEE Transactions on Geoscience and Remote Sensing 40*(11):2375–2383.

BIGGS J, T WRIGHT, Z LU, B PARSONS (2007): Multi-interferogram method for measuring interseismic deformation: Denali Fault, Alaska. *Geophysical Journal International 170*(3):1165–1179. DOI: 10.1111/j.1365-246X.2007.03415.x.

BING C, X SHAOJIAN, Z PING (2006): Estimation of InSAR baseline based on the Frequency Shift Theory. In: SHUNJUN W (Ed.), *Proceedings of the 2006 CIE International Conference on Radar, Shanghai, China, 16–19 Oct 2006*.

BOCK H, A JÄGGI, U MEYER, PNAM VISSER, J VAN DEN IJSSEL, T VAN HELLEPUTTE, M HEINZE, U HUGENTOBLER (2011): GPS-derived orbits for the GOCE satellite. *Journal of Geodesy 85*(11):807–818.

CAVALIÉ O, MP DOIN, C LASSERRE, P BRIOLE (2007): Ground motion measurement in the Lake Mead area, Nevada, by differential synthetic aperture radar interferometry time series analysis: Probing the lithosphere rheological structure. *Journal of Geophysical Research 112*:B03403. DOI: 10.1029/2006JB004344.

CHEN CW, HA ZEBKER (2001): Two-dimensional phase unwrapping with use of statistical models for cost functions in nonlinear optimization. *Journal of the Optical Society of America A 18*(2):338–351.

CLOSA J (1998): *The Influence of Orbit Precision in the Quality of ERS SAR Interferometric Data*. Technical document no. ES-TN-APP-APM-JC01, ESA.

COLESANTI C, A FERRETTI, F NOVALI, C PRATI, F ROCCA (2003): SAR Monitoring of Progressive and Seasonal Ground Deformation Using the Permanent Scatterers Technique. *IEEE Geoscience and Remote Sensing Letters 41*(7):1685–1701.

COVELLO F, F BATTAZZA, A COLETTA, G MANONI, G VALENTINI (2009): Cosmo-SkyMed Mission Status: Three out of Four Satellites in Orbit. In: *International Geoscience and Remote Sensing Symposium (IGARSS) 2009 Proceedings, vol. 2, Cape Town, South Africa, 12–17 July 2009*, 773–776. IEEE.

CSA (2010): *Characterizing Radarsat-1 Predicted Orbit Accuracies*. Operational Analysis Report RSCSA-OAR0183, CSA.

CSA (2012): *Earth-Observation satellites*. CSA, `http://www.asc-csa.gc.ca/eng/satellites/`, accessed on 16 Jan 2012.

CUMMING IG, FH WONG (2005): *Digital Processing of Synthetic Aperture Radar Data*. Artech House, Boston, USA.

CURLANDER JC, RN MCDONOUGH (1991): *Synthetic Aperture Radar: Systems and Signal Processing*. John Wiley & Sons, New York, USA.

DEMETS C, RG GORDON, DF ARGUS, S STEIN (1994): Effect of recent revisions to the geomagnetic reversal time scale on estimates of current plate motions. *Geophysical Research Letters 21*(20):2191–2194.

DING X, Z LI, J ZHU, G FENG, J LONG (2008): Atmospheric Effects on InSAR Measurements and Their Mitigation. *Sensors 8*(9):5426–5448.

DLR (2009): *TerraSAR-X Ground Segment Basic Product Specification Document*. Technical document no. TX-GS-DD-3302, DLR.

DLR (2013): *Tandem-L Science*. Microwaves and Radar Institute, DLR, `http://www.dlr.de/hr/en/desktop default.aspx/tabid-8113/`, accessed on 15 Feb 2013.

DOORNBOS E, R SCHARROO (2005): Improved ERS and ENVISAT Precise Orbit Determination. In: LACOSTE H, L OUWEHAND (Eds.), *Proceedings of the 2004 ENVISAT & ERS Symposium, Salzburg, Austria, 6–10 Sep 2004*, ESA SP-572.

DREWES H (2009): The Actual Plate Kinematic and Crustal Deformation Model APKIM2005 as Basis for a Non-Rotating ITRF. In: DREWES H (Ed.), *Geodetic Reference Frames, Munich, 9–14 Oct 2006*, IAG Symposia 134, 95–99. Springer, Berlin, Germany.

DUESMANN B, I BARAT (2007): *ERS-2 and ENVISAT orbit control – Current and future strategy: The impact on the interferometric baseline.* Presentation at the ESA Fringe 2007 Workshop, Frascati, Italy, 26–30 Nov 2007.

DUT (2009): *Delft Object-oriented Radar Interferometric Software: User's manual and technical documentation.* Delft University of Technology, Delft, The Netherlands. Version 4.02, http://doris.tudelft.nl.

EINEDER M, J HOLZNER (1999): Phase Unwrapping of Low Coherence Differential Interferograms. In: STEIN TI (Ed.), *International Geoscience and Remote Sensing Symposium (IGARSS) 1999 Proceedings, vol. 3, Hamburg, Germany, 28 June–2 July 1999*, 1727–1730. IEEE.

ELLIOT JR, J BIGGS, B PARSONS, TJ WRIGHT (2008): InSAR slip rate determination on the Altyn Tagh Fault, northern Tibet, in the presence of topographically correlated atmospheric delays. *Geophysical Research Letters 35*:L12309. DOI: 10.1029/2008GL033659.

ESA (2007a): *Envisat-1 Products Specifications. Volume 8: ASAR Products Specifications.* Product Document No. PO-RS-MDA-GS-2009, ESA.

ESA (2007b): *Information on ALOS PALSAR Products for ADEN Users.* Technical note no. ALOS-GSEG-EOPG-TN-07-0001, ESA.

ESA (2012): *Envisat operations.* ESA, http://www.esa.int/esaMI/Operations/SEMOZY8L6VE_0.html, accessed on 19 Jan 2012.

ESA (2013): *Sentinel-1: Mission Details.* ESA, http://earth.esa.int/web/guest/missions/esa-future-missions/sentinel-1, accessed on 15 Feb 2013.

FEIGL KL, F SARTI, H VADON, S MCCLUSKY, S ERGINTAV, P DURAND, R BÜRGMANN, A RIGO, D MASSONNET, R REILINGER (2002): Estimating Slip Distribution for the Izmit Mainshock from Coseismic GPS, ERS-1, RADARSAT, and SPOT Measurements. *Bulletin of the Seismological Society of America 92*(1):138–160.

FEIGL KL, CH THURBER (2009): A method for modelling radar interferograms without phase unwrapping: application to the M5 Fawnskin, California earthquake of 1992 December 4. *Geophysical Journal International 176*:491–504. DOI: 10.1111/j.1365-246X.2008.03881.x.

FERRETTI A, A FUMAGALLI, F NOVALI, C PRATI, F ROCCA, A RUCCI (2011): A New Algorithm for Processing Interferometric Data-Stacks: SqueeSAR. *IEEE Transactions on Geoscience and Remote Sensing 49*(9):3460–3470. DOI: 10.1109/TGRS.2011.2124465.

FERRETTI A, C PRATI, F ROCCA (2000): Nonlinear Subsidence Rate Estimation Using Permanent Scatterers in Differential SAR Interferometry. *IEEE Transactions on Geoscience and Remote Sensing 38*(5):2202–2212.

FERRETTI A, C PRATI, F ROCCA (2001): Permanent Scatterers in SAR Interferometry. *IEEE Transactions on Geoscience and Remote Sensing 39*(1):8–20.

FÖRSTNER W (1979): Ein Verfahren zur Schätzung von Varianz- und Kovarianzkomponenten. *Allgemeine Vermessungsnachrichten 86*(11-12):446–453.

GANTERT S, G RIEGLER, F TEUFEL, O LANG, L PETRAT, W KOPPE, J HERRMANN (2011): TerraSAR-X, TanDEM-X, TerraSAR-X2 and their Applications. In: *3rd Asia-Pacific International Conference on Synthetic Aperture Radar, Soeul, Korea, 26–30 Sep 2011.*

GATELLI F, A MONTI GUARNIERI, F PARIZZI, P PASQUALI, C PRATI, F ROCCA (1994): The Wavenumber Shift in SAR Interferometry. *IEEE Transactions on Geoscience and Remote Sensing 32*(4):855–865.

GEUDTNER D (1995): *Die interferometrische Verarbeitung von SAR-Daten.* Forschungsbericht 95-28, Deutsche Forschungsanstalt für Luft- und Raumfahrt (DLR), Oberpfaffenhofen, Germany.

GEUDTNER D, R TORRES, P SNOEIJ, M DAVIDSON (2011): *Sentinel-1 System Overview.* Presentation at the ESA Fringe 2011 Workshop, Frascati, Italy, 19–23 Sep 2011.

GEUDTNER D, PW VACHON, KE MATTAR, AL GRAY (1998): RADARSAT Repeat-Pass SAR Interferometry. In: STEIN TI (Ed.), *International Geoscience and Remote Sensing Symposium (IGARSS) 1998 Proceedings, vol. 3, Seattle, USA, 6–10 July 1998*, 1635–1637. IEEE.

GHIGLIA DC, MD PRITT (1998): *Two-Dimensional Phase Unwrapping*. John Wiley & Sons, New York, USA.

GOLDSTEIN RM, CL WERNER (1998): Radar Interferogram Filtering for Geophysical Applications. *Geophysical Research Letters 25*(21):4035–4038.

GÓMEZ B, MJ GONZÁLEZ, B BRÄUTIGAM, E VEGA, M GARCÍA, N CASAL, J DEL CASTILLO, JM CUERDA, N ALFARO, V ÁLVAREZ (2012): PAZ Misson: CALVAL Centre Activities. In: *9th European Conference on Synthetic Aperture Radar Proceedings, Nuremberg, Germany, 23–29 Apr 2012*.

HAMPEL F, E RONCHETTI, PJ ROUSSEEUW, W STAHEL (1986): *Robust Statistics*. John Wiley & Sons, New York, USA.

HANSSEN RF (2001): *Radar Interferometry – Data Interpretation and Error Analysis*. Kluwer, Dordrecht, The Netherlands.

HECK B (1981): Der Einfluß einzelner Beobachtungen auf das Ergebnis einer Ausgleichung und die Suche nach Ausreißern in den Beobachtungen. *Allgemeine Vermessungsnachrichten 88*(1):17–34.

HOOPER AJ (2008): A multi-temporal InSAR method incorporating both persistent scatterer and small baseline approaches. *Geophysical Research Letters 35*:L16302. DOI: 10.1029/2008GL034654.

HOOPER AJ (2010): A Statistical-Cost Approach to Unwrapping the Phase of InSAR Time Series. In: LACOSTE-FRANCIS H (Ed.), *Proceedings of 'Fringe 2009', Frascati, Italy, 30 Nov–4 Dec 2009*, ESA SP-677.

HOOPER AJ, P SEGALL, HA ZEBKER (2007): Persistent scatterer interferometric synthetic aperture radar for crustal deformation analysis, with application to Volcán Alcedo, Galápagos. *Journal of Geophysical Research 112*:B07407. DOI: 10.1029/2006JB004763.

HOOPER AJ, K SPAANS, D BEKAERT, M CARO CUENCA, M ARIKAN, A OYEN (2010): *Stamps/MTI Manual*. Delft University of Technology, Delft, The Netherlands. Version 3.2, `http://radar.tudelft.nl/~ahooper/stamps/`, accessed on 9 Mar 2012.

HOOPER AJ, HA ZEBKER (2007): Phase unwrapping in three dimensions with application to InSAR time series. *Journal of the Optical Society of America A 24*(9):2737–2747.

HOOPER AJ, HA ZEBKER, P SEGALL, BM KAMPES (2004): A new method for measuring deformation on volcanoes and other natural terrains using InSAR persistent scatterers. *Geophysical Research Letters 31*:L23611. DOI: 10.1029/2004GL021737.

HUBER PJ (1981): *Robust Statistics*. John Wiley & Sons, New York, USA.

IGN (2007): *ITRS and WGS84*. IGN, `ftp://itrf.ensg.ign.fr/pub/itrf/WGS84.TXT`, accessed on 1 Oct 2012.

IMPAGNATIELLO F, R BERTONI, F CALTAGIRONE (1998): The SkyMed/COSMO System : SAR Payload Characteristics. In: STEIN TI (Ed.), *International Geoscience and Remote Sensing Symposium (IGARSS) 1998 Proceedings, vol. 2, Seattle, USA, 6–10 July 1998*, 689–691. IEEE.

JÄGER R, T MÜLLER, H SALER, R SCHWÄBLE (2006): *Klassische und robuste Ausgleichungsverfahren*. Wichmann, Heidelberg, Germany.

JAXA (2008): *ALOS Data Users Handbook, Revision C*. JAXA.

JAXA (2012): *ALOS-2: The Advanced Land Observing Satellite-2*. Brochure, JAXA, `http://www.jaxa.jp/pr/brochure/pdf/04/sat29.pdf`, accessed on 22 Jan 2012.

JAYLES C, B NHUN-FAT, C TOURAIN (2006): DORIS: system description and control of the signal integrity. *Journal of Geodesy 80*(8-11):457–472.

JÓNSSON S (2002): *Modeling Volcano and Earthquake Deformation from Satellite Radar Interferometric Observations*. PhD thesis, Stanford University, Stanford, USA.

JURISCH R, G KAMPMANN (1998): Vermittelnde Ausgleichungsrechnung mit balancierten Beobachtungen – erste Schritte zu einem neuen Ansatz. *Zeitschrift für Vermessungswesen 123*(3):87–92.

KAMPES BM (2006): *Radar Interferometry – Persistent Scatterer Technique*. Springer, Dordrecht, The Netherlands.

KAMPES BM, RF HANSSEN (2004): Ambiguity Resolution for Permanent Scatterer Interferometry. *IEEE Transactions on Geoscience and Remote Sensing 42*(11):2446–2453.

KAMPES BM, RF HANSSEN, Z PERSKI (2004): Radar Interferometry with Public Domain Tools. In: LACOSTE H (Ed.), *Proceedings of FRINGE 2003 Workshop, Frascati, Italy, 1–5 Dec 2003*, ESA SP-550.

KANKAKU Y, Y OSAWA, S SUZUKI, T WATANABE (2009): The Overview of the L-band SAR Onboard ALOS-2. In: *Progress in Electromagnetics Research Symposium, Moscow, Russia, 18–21 Aug 2009*.

KETELAAR VBH (2009): *Satellite radar interferometry – subsidence monitoring techniques*. Springer, Berlin, Germany.

KNEDLIK S, O LOFFELD, A HEIN, C ARNDT (1999): A Novel Approach to Accurate Baseline Estimation. In: STEIN TI (Ed.), *International Geoscience and Remote Sensing Symposium (IGARSS) 1999 Proceedings, vol. 1, Hamburg, Germany, 28 June–2 July 1999*. IEEE.

KNOSPE SHG, S JÓNSSON (2010): Covariance Estimation for dInSAR Surface Deformation Measurements in the Presence of Anisotropic Atmospheric Noise. *IEEE Transactions on Geoscience and Remote Sensing 48*(4):2057–2065. DOI: 10.1109/TGRS.2009.2033937.

KOCH KR (1996): Robuste Parameterschätzung. *Allgemeine Vermessungsnachrichten 103*(1):1–18.

KOCH KR (1999): *Parameter estimation and hypothesis testing in linear models*, 2nd edition. Dümmler, Bonn, Germany.

KOHLHASE AO, KL FEIGL, D MASSONNET (2003): Applying differential InSAR to orbital dynamics: a new approach for estimating ERS trajectories. *Journal of Geodesy 77*(9):493–502.

KREYSZIG E (1977): *Statistische Methoden und ihre Anwendungen*, 6th edition. Vandenhoeck & Ruprecht, Göttingen, Germany.

KRIEGER G, A MOREIRA, H FIEDLER, I HAJNSEK, M WERNER, M YOUNIS, M ZINK (2007): TanDEM-X: A Satellite Formation for High-Resolution SAR Interferometry. *IEEE Transactions on Geoscience and Remote Sensing 45*(11):3317–3341.

LAUR H (2011): *SAR Interferometry opportunities with the European Space Agency*. Presentation at the ESA Fringe 2011 Workshop, Frascati, Italy, 19–23 Sep 2011.

VAN LEIJEN FJ, RF HANSSEN (2007): Persistant Scatterer Interferometry using Adaptive Deformation Models. In: LACOSTE H, L OUWEHAND (Eds.), *Proceedings of the Envisat Symposium 2007, Montreux, Switzerland, 23–27 Apr 2007*, ESA SP-636.

VAN LEIJEN FJ, RF HANSSEN, PS MARINKOVIC, BM KAMPES (2006): Spatio-temporal Phase Unwrapping using Integer Least-Squares. In: LACOSTE H (Ed.), *Proceedings of Fringe 2005 Workshop, Frascati, Italy, 28 Nov– 2 Dec 2005*, ESA SP-610.

LIU S, RF HANSSEN, ÁGNES MIKA (2010): Feasibility of retrieving Spatial Variations of Atmospheric Phase Screen at Epochs of SAR Acquisitions from SAR interferometry. In: LACOSTE-FRANCIS H (Ed.), *Proceedings of 'Fringe 2009', Frascati, Italy, 30 Nov–4 Dec 2009*, ESA SP-677.

LUNDGREN P, EA HETLAND, Z LIU, EJ FIELDING (2009): Southern San Andreas-San Jacinto fault system slip rates estimated from earthquake cycle models constrained by GPS and interferometric synthetic aperture radar observations. *Journal of Geophysical Research 114*:B02403. DOI: 10.1029/2008JB005996.

MARINKOVIC P, Y LARSEN (2013): On the long term stability of ASAR LO frequency – empirical study of 10 years of data. *IEEE Transactions on Geoscience and Remote Sensing*. Submitted.

MASSMANN FH, F FLECHTNER, JC RAIMONDO, C REIGBER (1997): Impact of PRARE on ERS-2 POD. *Advances in Space Research 19*(11):1645–1648.

MASSONNET D, KL FEIGL (1998): Radar Interferometry and its applications to changes in the Earth's surface. *Reviews of Geophysics 36*(4):441–500.

MASSONNET D, H VADON (1995): ERS-1 Internal Clock Drift measured by Interferometry. *IEEE Transactions on Geoscience and Remote Sensing 33*(2):401–408.

MDA (2009): *Radarsat-2 Product Description.* Technical document no. RN-SP-52-1238, MDA.

MDA (2013): *About RADARSAT-2.* MDA, `http://gs.mdacorporation.com/SatelliteData/Radarsat2/About.aspx`, accessed on 15 Feb 2013.

MEYER FJ, R BAMLER, N JAKOWSKI, T FRITZ (2006): The Potential of Low-Frequency SAR Systems for Mapping Ionospheric TEC Distributions. *IEEE Geoscience and Remote Sensing Letters 3*(4):560–564. DOI: 10.1109/LGRS.2006.882148.

MEZZASOMA S, A GALLON, F IMPAGNATIELLO, G ANGINO, S FAGIOLI, A CAPUZI, F CALTAGIRONE, R LEONARDI, U ZILIOTTO (2008): COSMO-SkyMed System Commissioning: End-to-End System Performance Verification. In: *Radar Conference 2008, Rome, Italy, 26–30 May 2008.* IEEE.

MIRANDA N, B ROSICH (2011): *Sentinel-1 Processor and Core Products.* Presentation at the ESA Fringe 2011 Workshop, Frascati, Italy, 19-23 Sep 2011.

MONTENBRUCK O, E GILL (2000): *Satellite Orbits.* Springer, Berlin, Germany.

MONTENBRUCK O, T VAN HELLEPUTTE, R KROES, E GILL (2004): Reduced dynamic orbit determination using GPS code and carrier measurements. *Aerospace Science and Technology 9*(3):261–271.

MONTI GUARNIERI A, P BIANCARDI, D D'ARIA, F ROCCA (2000): Accurate and robust baseline estimation. In: *FRINGE '99: Advancing ERS SAR Interferometry from Applications towards Operations, Liège, Belgium, 10–12 Nov 1999,* ESA SP-478.

MOREIRA A (2000): *Radar mit synthetischer Apertur.* Habiliation treatise, Fakultät für Elektrotechnik und Informationstechnik, Universität Karlsruhe (TH), Germany.

MOREIRA A, G KRIEGER, M YOUNIS, I HAJNSEK, KP PAPATHANASSIOU, M EINEDER, F DE ZAN (2011): Tandem-L: A Mission Proposal for Monitoring Dynamic Earth Processes. In: *International Geoscience and Remote Sensing Symposium (IGARSS) 2011 Proceedings, Vancouver, Canada, 24–29 July 2011,* 1385–1388. IEEE.

MORENA LC, KV JAMES, J BECK (2004): An introduction to the RADARSAT-2 mission. *Canadian Journal of Remote Sensing 30*(3):221–234.

MURAKAMI M, M TOBITA, S FUJIWARA, T SAITO, H MASAHARU (1996): Coseismic crustal deformations of 1994 Northridge, California, earthquake detected by interferometric JERS 1 synthetic aperture radar. *Journal of Geophysical Research 101*(B4):8605–8614.

NAKAMURA R, S NAKAMURA, N KUDO, S KATAGIRI (2007): Precise Orbit Determination for ALOS. In: *20th International Symposium on Space Flight Dynamics, Annapolis, USA, 24–28 Sep 2007.*

NIEMEIER W (2008): *Ausgleichungsrechnung,* 2nd edition. Walter de Gruyter, Berlin.

NITTI DO, RF HANSSEN, A REFICE, F BOVENGA, R NUTRICATO (2011): Impact of DEM-Assisted Coregistration on High-Resolution SAR Interferometry. *IEEE Transactions on Geoscience and Remote Sensing 49*(3):1127–1143.

OTTEN M, JM DOW (2005): ENVISAT Precise Orbit Determination. In: LACOSTE H, L OUWEHAND (Eds.), *Proceedings of the 2004 ENVISAT & ERS Symposium, Salzburg, Austria, 6–10 Sep 2004,* ESA SP-572.

OTTEN M, PNAM VISSER, FH MASSMANN, S RUDENKO, R SCHARROO (2011): *ERS-1 and ERS-2 Orbit Validation Report.* Technical report, ESA.

PEPE A, P BERARDINO, M BONANO, LD EUILLADES, R LANARI, E SANSOSTI (2011): SBAS-Based Satellite Orbit Correction for the Generation of DInSAR Time-Series: Application to RADARSAT-1 Data. *IEEE Transactions on Geoscience and Remote Sensing 49*(12):5150–5165.

PETIT G, B LUZUM (2010): *IERS Conventions*, IERS Technical Note 36. Verlag des Bundesamts für Kartographie und Geodäsie, Frankfurt am Main, Germany.

POPE AJ (1976): *The Statistics of Residuals and the Detection of Outliers*. NOAA Technical Report NOS 65 NGS 1, US Department of Commerce, Rockville, Maryland, USA.

PRATI C, F ROCCA (1990): Limits to the Resolution of Elevation Maps from Stereo SAR Images. *International Journal of Remote Sensing 11*(12):2215–2235.

PRATI C, F ROCCA, A MONTI GUARNIERI (1993): SAR Interferometry Experiments with ERS-1. In: KALDEICH B (Ed.), *Proceedings of the First ERS-1 Symposium: Space at the Service of Our Environment, Cannes, France, 4–6 Nov 1992*, 211–218.

RICHARDS JA (2009): *Remote Sensing with Imaging Radar*. Springer, Berlin, Germany.

ROSEN PA, S HENSLEY, FK LI, SN MADSEN, E RODRÍGUEZ, RM GOLDSTEIN (2000): Synthetic Aperture Radar Interferometry. *Proceedings of the IEEE 88*(3):333–382.

ROUSSEEUW PJ, AM LEROY (1987): *Robust Regression and Outlier Detection*. John Wiley & Sons, New York, USA.

RSI (2000): *Radarsat Data Products Specifications*. Technical document no. RSI-GS-026, Radarsat International.

SAUNIER S, P GORYL, M BOUVET, R SANTER, A GRUEN, K WOLF, F VIALLEFONT (2007): *The contribution of the European Space Agency to the commissioning phase of the PRISM & AVNIR-2 instruments onboard ALOS (JAXA)*. Presentation at the IEEE International Geoscience and Remote Sensing Symposium (IGARSS), Barcelona, Spain, 23–27 July 2007.

SCHARROO R (2002): *A Decade of ERS Satellite Orbits and Altimetry*. Delft University Press, Delft, The Netherlands.

SCHARROO R, PNAM VISSER (1998): Precise orbit determination and gravity field improvement for the ERS satellites. *Journal of Geophysical Research 103*(C4):8113–8127.

SCHMIDT DA, R BÜRGMANN (2003): Time-dependent land uplift and subsidence in the Santa Clara valley, California, from a large interferometric synthetic aperture radar data set. *Journal of Geophysical Research 108*(B9):2416.

SCHUBERT A, M JEHLE, D SMALL, E MEIER (2012): Mitigation of atmospheric perturbations and solid Earth movements in a TerraSAR-X time-series. *Journal of Geodesy 86*(4):257–270. DOI: 10.1007/s00190-011-0515-6.

SCHWÄBISCH M, D GEUDTNER (1995): Improvement of phase and coherence map quality using azimuth prefiltering: examples from ERS-1 and X-SAR. In: *International Geoscience and Remote Sensing Symposium (IGARSS) 1995 Proceedings, vol. 1, Florence, Italy, 10–14 July 1995*, 205–207. IEEE.

SELLA GF, TH DIXON, A MAO (2002): REVEL: A model for Recent plate velocities from space geodesy. *Journal of Geophysical Research 107*(B4):2081. DOI: 10.1029/2000JB000033.

SHIRZAEI M, T WALTER (2011): Estimating the Effect of Satellite Orbital Error Using Wavelet-Based Robust Regression Applied to InSAR Deformation Data. *IEEE Transactions on Geoscience and Remote Sensing 49*(11):4600–4605. DOI: 10.1109/TGRS.2011.2143419.

SIMONETTO E (2008): DINSAR Experiments using a Free Processing Chain. In: BRUZZONE L, C NOTARNICOLA, F POSA (Eds.), *Image and Signal Processing for Remote Sensing XIV, Cardiff, UK, 15–18 Sep 2008*, Proceedings of SPIE 7109, 71091G. DOI: 10.1117/12.802508.

SINGH K, N STUSSI, KL KEONG, L HOCK (1997): Baseline Estimation In Interferometric SAR. In: STEIN TI (Ed.), *International Geoscience and Remote Sensing Symposium (IGARSS) 1997 Proceedings, vol. 1, Singapore, 3–8 Aug 1997*, 454–456.

SMALL D, CL WERNER, D NÜESCH (1993): Baseline Modelling for ERS-1 SAR Interferometry. In: *International Geoscience and Remote Sensing Symposium (IGARSS) 1993 Proceedings, Tokyo, Japan, 18–21 Aug 1993*, 1204–1206. IEEE.

TEUNISSEN PJG (2000): *Adjustment theory.* Delft University Press, Delft, The Netherlands.

THOMPSON AA (2010): Innovative Capabilities of the RADARSAT Constellation Mission. In: *8th European Conference on Synthetic Aperture Radar Proceedings, Aachen, Germany, 7–10 June 2010.*

VACHON PW, D GEUDTNER, AL GRAY, R TOUZI (1995): ERS-1 Synthetic Aperture Radar Repeat-Pass Interferometry Studies: Implications for Radarsat. *Canadian Journal of Remote Sensing 21*(4):441–454.

VISSER PNAM, R SCHARROO, R FLOBERGHAGEN, BAC AMBROSIUS (1997): Impact of PRARE on ERS-2 Orbit Determination. In: GUYENNE TD (Ed.), *Proceedings of the 12th International Symposium on Space Flight Dynamics, Darmstadt, Germany, 2–6 June 1997*, ESA SP-403.

WEBSTER R, MA OLIVER (2007): *Geostatistics for Environmental Scientists.* John Wiley & Sons, Chichester, UK.

WEGMÜLLER U, CL WERNER, T STROZZI, A WIESMANN (2006): Ionospheric Electron Concentration Effects on SAR and INSAR. 3714–3717. IEEE.

WERNER CL, S HENSLEY, RM GOLDSTEIN, PA ROSEN, HA ZEBKER (1993): Techniques and Applications of SAR Interferometry for ERS-1: Topography, Change Detection and Slope Measurement. In: KALDEICH B (Ed.), *Proceedings of the First ERS-1 Symposium: Space at the Service of Our Environment, Cannes, France, 4–6 Nov 1992*, 205–210. ESA.

WERNINGHAUS R, S BUCKREUSS (2010): The TerraSAR-X Mission and System Design. *IEEE Transactions on Geoscience and Remote Sensing 48*(2):606–614.

WICKI F (1998): *Robuste Schätzverfahren für die Parameterschätzung in geodätischen Netzen.* Mitteilungen no. 67, Institut für Geodäsie und Photogrammetrie, ETH Zürich, Zürich, Switzerland.

WRIGHT T, M GARTHWAITE, HS JUNG, A SHEPHERD (2011): *How accurately can current and future InSAR missions map tectonic strain?* Presentation at the ESA Fringe 2011 Workshop, Frascati, Italy, 19–23 Sep 2011.

XIA Y (2010): Synthetic Aperture Radar Interferometry. In: XU G (Ed.), *Sciences of Geodesy I: Advances and Future Directions*, chapter 11, 415–474. Springer, Berlin, Germany.

YOON YT, M EINEDER, N YAGUE-MARTINEZ, O MONTENBRUCK (2009): TerraSAR-X Precise Trajectory Estimation and Quality Assessment. *IEEE Transactions on Geoscience and Remote Sensing 47*(6):1859–1868.

ZEBKER HA, F AMELUNG, S JÓNSSON (2000): Remote Sensing of Volcano Surface and Internal Processes Using Radar Interferometry. In: MOUGINIS-MARK PJ, JA CRISP, JH FINK (Eds.), *Remote Sensing of Active Volcanism*, Geophysical Monograph 116, 179–205. AGU, Washington D. C., USA.

ZEBKER HA, RM GOLDSTEIN (1986): Topographic Mapping from Interferometric Synthetic Aperture Radar Observations. *Journal of Geophysical Research 91*(B9):4993–4999.

ZEBKER HA, PA ROSEN, RM GOLDSTEIN, AK GABRIEL, CL WERNER (1994a): On the derivation of coseismic displacement fields using differential radar interferometry: The Landers earthquake. *Journal of Geophysical Research 99*(B10):19617–19634.

ZEBKER HA, PA ROSEN, S HENSLEY (1997): Atmospheric effects in interferometric synthetic aperture radar surface deformation and topographic maps. *Journal of Geophysical Research 102*(B4):7547–7563.

ZEBKER HA, J VILLASENOR (1992): Decorrelation in Interferometric Radar Echoes. *IEEE Transactions on Geoscience and Remote Sensing 30*(5):950–959.

ZEBKER HA, CL WERNER, PA ROSEN, S HENSLEY (1994b): Accuracy of Topographic Maps Derived from ERS-1 Interferometric Radar. *IEEE Transactions on Geoscience and Remote Sensing 32*(4):823–836.

ZHANG K, AHM NG, X LI, HC CHANG, L GE, C RIZOS (2009): A New Approach to improve the Accuracy of Baseline Estimation for Spaceborne Radar Interferometry. In: *International Geoscience and Remote Sensing Symposium (IGARSS) 2009 Proceedings, vol. 5, Cape Town, South Africa, 12–17 July 2009*. IEEE.

ZHAO HL, JH FAN, XF GUO (2010): A method for InSAR baseline refinement and its application. In: LUO Q (Ed.), *Second IITA International Conference on Geoscience and Remote Sensing Proceedings, vol. 2, Qingdao, China, 28–31 Aug 2010*, 161–164. IEEE.

ZINK M, M BARTUSCH, D MILLER (2011): TanDEM-X Mission Status. In: *International Geoscience and Remote Sensing Symposium (IGARSS) 2011 Proceedings, Vancouver, Canada, 24–29 July 2011*, 2290–2293. IEEE.

List of Symbols

Timescales

T long-term acquisition time ($\sim$ d...a)
t azimuth acquisition time (slow time, $\sim$ s)
τ two-way ranging time (fast time, $\sim$ ms)

General conventions

$\mathbf{X}$ matrix
$\mathbf{x}$ vector in an algebraic sense
$\mathbf{X}^T, \mathbf{x}^T$ transpose of a matrix or a vector
$\mathbf{X}^{-1}$ Cayley inverse of a square matrix
$\mathbf{X}^*$ matrix in the model without elimination of phase offset parameters
$\mathbf{X} \otimes \mathbf{Y}$ Kronecker product
$\vec{x}$ 3×1 vector in a geometric sense
$\langle \vec{x}, \vec{y} \rangle$ dot product, inner product
$\vec{x} \times \vec{y}$ cross product, vector product
x scalar
x^* complex conjugate
$\hat{x}$ estimator, estimated quantity
$\tilde{x}$ biased quantity
$\dot{x}$ derivative with respect to the azimuth acquisition time t
x' derivative with respect to the long-term acquisition time T
Δ difference operator
δ prefix for a residual or error quantity

Operators

$\mathrm{D}\{\cdot\}$ dispersion
$\mathrm{E}\{\cdot\}$ expectation
$\mathcal{F}\{\cdot\}$ Fourier transform
$\mathrm{tr}\{\cdot\}$ trace

Specific symbols

List of Acronyms

AGU	American Geophysical Union
APS	Atmospheric Phase Screen
ASAR	Advanced Synthetic Aperture Radar (Envisat)
ASI	Agenzia Spaziale Italiana (Italian Space Agency)
ASTER	Advanced Spaceborne Thermal Emission and Reflection Radiometer
CNES	Centre National d'Etudes Spatiales (National Centre for Space Studies, France)
CSA	Canadian Space Agency
DEM	Digital Elevation Model
DInSAR	Differential InSAR
DLR	Deutsches Zentrum für Luft- und Raumfahrt (German Aerospace Center)
DOF	Degree of Freedom
DORIS	Delft Object-oriented Radar Interferometric Software
	or: Doppler Orbitography and Radiopositioning Integrated by Satellite
DUT	Delft University of Technology (the Netherlands)
Envisat	Environmental Satellite
ERS	European Remote Sensing Satellite
ESA	European Space Agency
ESOC	European Space Operations Centre
GCP	Ground Control Point
GNSS	Global Navigation Satellite System
GPS	Global Positioning System
GRACE	Gravity Recovery and Climate Experiment
IEEE	Institute of Electrical and Electronics Engineers
IERS	International Earth Rotation and Reference Systems Service
InSAR	Interferometric SAR, SAR Interferometry
IS2	Image Swath 2 (Envisat)
ITRF	International Terrestrial Reference Frame
ITRS	International Terrestrial Reference System
IWS	Interferometric Wide Swath (Sentinel-1)
JAXA	Japan Aerospace Exploration Agency
JERS	Japanese Earth Resources Satellite
KIT	Karlsruhe Institute of Technology (Germany)
LOS	Line of Sight
MDA	MacDonald, Dettwiler and Associates Ltd. (Canada)
MERIS	Medium Resolution Imaging Spectroradiometer (Envisat)
MODIS	Moderate Resolution Imaging Spectroradiometer
NASDA	National Space Development Agency (Japan, until 2003)
NCG	Nederlandse Commissie voor Geodesie (Netherlands Geodetic Commission)
NNR-NUVEL	No Net Rotation – Northwestern University Velocity Model
NOAA	National Oceanic and Atmospheric Administration (USA)
PALSAR	Phased Array type L-band SAR (ALOS)
PDF	Probability Density Function

POD	Precise Orbit Determination
PRARE	Precise Range And Range-Rate Equipment (ERS-2)
PRF	Pulse Repetition Frequency
PS	Persistent Scatterer
PSC	Persistent Scatterer Candidate
Radar	Radio Detection and Ranging
RMS	Root Mean Square
RSR	Range Sampling Rate
SAR	Synthetic Aperture Radar
SBAS	Small Baseline Subsets
SLC	Single Look Complex
SLR	Satellite Laser Ranging
SNAPHU	Statistical-Cost Network-Flow Algorithm for Phase Unwrapping
SRTM	Shuttle Radar Topography Mission
SWST	Sampling Window Start Time
TEC	Total Electron Content
TECU	TEC Unit
TRF	Terrestrial Reference Frame
UK	United Kingdom of Great Britain and Northern Ireland
USA	United States of America
VLBI	Very Long Baseline Interferometry

Acknowledgements

I would like to thank my principal advisor, Prof. Dr.-Ing. Dr.-Ing. E. h. Günter Schmitt, for his supervision and his commitment to support my research throughout all stages of thesis preparation and completion. I am further indebted to Prof. dr. ir. Ramon Hanssen for his guidance, vision and patience as well as his three-fold welcome at Delft University of Technology. I also want to acknowledge Prof. Dr.-Ing. Dr. h. c. Bernhard Heck for his interest, his involvement and for giving valuable feedback.

I wish to express my sincere recognition to my colleagues and friends at the Geodetic Institute Karlsruhe (GIK) and the Institute of Photogrammetry and Remote Sensing (IPF, Karlsruhe) for their good collaboration and for contributing to a pleasant working atmosphere. I particularly appreciated the lively technical and non-technical discussions during tea breaks at the branch office "Post". Special thanks go to Dr. rer. nat. Malte Westerhaus who was a great help when getting started and was willing to listen and give advice whenever I approached him.

I am grateful to the members of the InSAR working group in Delft for their friendly welcome and many inspiring discussions. It was a pleasure to temporarily work in their topic-focussed environment, which always stimulated my motivation and made me return to Karlsruhe with fresh ideas.

I would like to thank Prof. Dr. rer. nat. Martin Breunig for giving me the opportunity to finish my dissertation after my principal advisor retired. I further want to acknowledge the Karlsruhe House of Young Scientists (KHYS) and the Graduate School for Climate and Environment (GRACE) for funding my research stays in Delft.

Finally, I owe my deepest gratitude to my dear parents for their unconditional support, sincere faith and continuous encouragement.

SCHRIFTENREIHE DES STUDIENGANGS GEODÄSIE UND GEOINFORMATIK

Die Bände sind unter www.ksp.kit.edu als PDF frei verfügbar oder als Druckausgabe bestellbar.
(ISSN 1612-9733)

2003/1   **Daniel Koenig, Kurt Seitz**
Numerische Integration von Satellitenbahnen unter Berücksichtigung der Anisotropie
des Gravitationsfeldes der Erde. 2003
ISBN 3-937300-00-7

2004/1   **Marcus Ulrich Schmidt**
Objektorientierte Modellierung zur geodätischen Deformationsanalyse. 2004
ISBN 3-937300-06-6

2005/1   **Stephan Kupferer**
Anwendung der Total-Least-Squares-Technik bei geodätischen Problemstellungen. 2005
ISBN 3-937300-67-8

2007/1   **Andreas Knöpfler, Michael Mayer, André Nuckelt, Bernhard Heck, Günter Schmitt**
Untersuchungen zum Einfluss von Antennenkalibrierwerten auf die Prozessierung
regionaler GPS-Netze. 2007
ISBN 978-3-86644-110-1

2007/2   **Xiaoguang Luo, Miachel Mayer, Bernhard Heck**
Bestimmung von hochauflösenden Wasserdampffeldern unter
Berücksichtigung von GNSS-Doppeldifferenzresiduen. 2007
ISBN 978-3-86644-115-6

2007/3   **Michael Mürle**
Aufbau eines Wertermittlungsinformationssystems. 2007
ISBN 978-3-86644-116-3

2007/4   **Heinrich Derenbach, Michael Illner, Günter Schmitt, Martin Vetter, Siegrfied Vielsack**
Ausgleichsrechnung – Theorie und aktuelle Anwendungen aus der Vermessungspraxis. 2007
ISBN 978-3-86644-124-8

2007/5   **André Nuckelt**
Dreidimensionale Plattenkinematik. 2007
ISBN 978-3-86644-152-1

2007/6   **Hermann Bähr, Zuheir Altamimi, Bernhard Heck**
Variance Component Estimation for Combination of Terrestrial Reference Frames. 2007
ISBN 978-3-86644-206-1

2009/1   **Cornelia Eschelbach**
Refraktionskorrekturbestimmung durch Modellierung des Impuls-
und Wärmeflusses in der Rauhigkeitsschicht. 2009
ISBN 978-3-86644-307-5

SCHRIFTENREIHE DES STUDIENGANGS GEODÄSIE UND GEOINFORMATIK

Die Bände sind unter www.ksp.kit.edu als PDF frei verfügbar oder als Druckausgabe bestellbar.
(ISSN 1612-9733)

2009/2   Bernhard Heck, Michael Mayer (Hrsg.)
Geodätische Woche 2009
22.–24. September 2009, Messe Karlsruhe, Rheinstetten im Rahmen der INTERGEO –
Kongress und Fachmesse für Geodäsie, Geoinformation und Landmanagement. Abstracts. 2009
ISBN 978-3-86644-411-9

2010/1   Thomas Grombein, Kurt Seitz, Bernhard Heck
Untersuchungen zur effizienten Berechnung topographischer Effekte auf
den Gradiententensor am Fallbeispiel der Satellitengradiometriemission GOCE. 2010
ISBN 978-3-86644-510-9 / KIT Scientific Reports 7547

**2010/2   Thomas Fuhrmann, Andreas Knöpfler, Michael Mayer,
         Xiaoguang Luo, Bernhard Heck**
Zur GNSS-basierten Bestimmung des atmosphärischen Wasserdampfgehalts mittels Precise
Point Positioning. 2010
ISBN 978-3-86644-539-0 / KIT Scientific Reports 7561

2010/3   Geodätisches Institut (Hrsg.)
Vernetzt und ausgeglichen
Festschrift zur Verabschiedung von Prof. Dr.-Ing. habil. Dr.-Ing. E.h. Günter Schmitt. 2010
ISBN 978-3-86644-576-5

2012/1   Michael Mayer, Claudia P. Krueger, Bernhard Heck (Hrsg.)
Highly Precise Positioning and Height Determination using GPS. Results of a
PROBRAL project by Universidade Federal do Paraná (UFPR, Curitiba, Brazil)
and Karlsruhe Institut für Technologie. 2012
KIT Scientific Reports 7604

2013/1   Melanie Müßle, Bernhard Heck, Kurt Seitz, Thomas Grombein
Untersuchungen zur planaren Approximation im Geodätschen Randwertproblem. 2013
ISBN 978-3-7315-0095-7 / KIT Scientific Reports 7652

2013/2   Hermann Bähr
Orbital Effects in Spaceborne Synthetic Aperture Radar Interferometry. 2013
ISBN 978-3-7315-0134-3